Introduction to
Applied Mathematics

Fourth Edition

Alan Parks
Lawrence University
Appleton, Wisconsin

♠ This book is the printed fourth edition of a text for the two-term Applied Calculus course at Lawrence University. The original project that grew into this text was begun during the summer of 2011.

Contents

Introduction

Mathematics has suffered from its own success. Mathematical techniques have yielded insights into virtually every area of quantitative work, and those techniques have grown so far and so fast and for so long that it is hard to keep track of them, let alone understand them comprehensively. Furthermore, the theoretical underpinnings of the subject have attracted focussed, inventive, and sustained thought over the past 2300 years, at least. Mathematics is very broad and very deep. How do you introduce its connections to other disciplines? We feel rather strongly that the answer involves introducing the reasoning that produces those connections, and then by teaching the *analytical* and *numerical* work that proffers insights into other disciplines.

Analytical work is that done by exact techniques in the *calculus* and *linear algebra*. The calculus was developed between about 1660 and the late 1800's; it centers on the idea of the *limit* – the computation of quantities by successive approximations. Linear algebra uses matrices to represent sequences and to solve linear equations. *Numerical* calculations are those done on a computer or calculator, where the numbers considered are, in general, approximate. These approaches will be intertwined. Our overall goals are (1) to provide you with analytical tools used both to express the main ideas in quantitative disciplines and to make deductions from that theory; (2) to provide you with numerical tools that will enable you to investigate applied problems directly and in detail.

We also have important secondary goals: to increase your facility with the most common and important mathematical functions; to increase your ability to read and understand analytic descriptions of variables and their relations; to increase your confidence in computation.

To be more specific about subject matter: during the first term (Applied Calculus I), we will study recursive sequences, the derivative, and linear optimization problems. The second term (Applied Calculus II) will involve integration, linear algebra, multivariate derivatives, and non-linear optimization problems.

This text will be coordinated with class work and homework; above all, this book is meant to be read carefully. Many students reading mathematics seriously are surprised at how long it takes to wade through a short section of text. Perhaps that's because lines of reasoning that seemed clear when you scanned them quickly dissolve when you try to reproduce or apply them. And the ability to reproduce and apply what you have read is the principal test of mathematical understanding.

There are many problems presented during the exposition: labeled with the letter T followed by the chapter number, then a period, and then the problem number. There is a separate problems document with a great number of exercises from which class examples and homework will be chosen.

Even though this book is intended to be self-contained, you are welcome to consult other books if you find them helpful. Our bibliography (on p.237) was taken, for the most part, from Lawrence University's Seeley G. Mudd Library. The books [1], [2], and [3], for example, are widely used textbooks. Indeed, there are an enormous number of texts for courses such as this one, and the student interested in additional problems or alternative approaches will have no trouble finding help from libraries and online. You might be helped occaisionally by a textbook for the general calculus, such as [13].

We hope you will be drawn into applied mathematics in a way that transcends the memorization of algorithmic calculation – we hope you will see interesting corridors to other disciplines, and, more than anything else, that you will gain confidence in recognizing applied problems and in addressing them effectively. We have intended for the level of this material to increase somewhat as the course moves along. We have observed that the reader's facility in reading mathematics increases correspondingly.

CHAPTER 1

The Exponential Function

1. Exponentials and Logarithms

An *exponential function*[1] is a function such as 2^x or 3^x or $(1/2)^x$, where the variable x is in the *exponent*. It turns out that we can deal with all exponential functions by understanding a particular exponential function, where the base is the special number e. The number e is hard to define; here is a very famous formula for it.

$$e = 2 + \frac{1}{2} + \frac{1}{3 \cdot 2} + \frac{1}{4 \cdot 3 \cdot 2} + \frac{1}{5 \cdot 4 \cdot 3 \cdot 2} + \frac{1}{6 \cdot 5 \cdot 4 \cdot 3 \cdot 2} + \cdots$$

The formula goes on forever, and you might recognize the denominators as *factorials*. Anyway, we can use this formula to get a good *approximation* to the number e.

$$e \approx 2.718281828459\ldots$$

There is no obvious pattern in the decimals, and, it turns out, e is an *irrational number* – that is to say it can't be written as a ratio of integers. We almost always just use the letter e to stand for the number and leave it like that – the way we use π for the circle number. Calculators use an approximation like the one we just gave.

The function e^x is called the *exponential function*; it is the most important function in both applied and theoretical mathematics. The definite article *the* in the name *the exponential function* identifies the use of the base e. When

---

[1]Most italicized terms will be found in the Index at the end of the book. We have indexed both terms and symbolic expressions.

another base is used, as in 2^x, the function is *an* exponential function. We will see that we can use *the* exponential function e^x to understand all the other exponential functions. There are very good reasons for using the base e, as we will see later in the course. For now, we want to review properties of the exponential function and its associated logarithm. Their graphs are given on p.13.

Properties of e^x

    (1) e^a is a positive real number for every real number a

    (2) $e^0 = 1$

    (3) $e^{a+b} = e^a \cdot e^b$ for all real numbers a, b

    (4) $(e^a)^b = e^{a \cdot b}$ for all real numbers a, b

The listed properties are a matter of algebra when a and b are *rational numbers* – when a, b can be written as ratios of integers. We will use these properties no matter what a, b are. Here is a consequence; make sure you see which properties (1)-(4) are being used in each step.

$$e^a \cdot e^{-a} = e^{a-a} = e^0 = 1 \quad \text{so that} \quad e^{-a} = \frac{1}{e^a}$$

This shows that we can switch an exponential from the numerator to the denominator by negating the exponent. This works for moving from the denominator to the numerator as well:

$$\frac{1}{e^{-3}} = e^3 \quad \text{and} \quad \frac{1}{e^4} = e^{-4}$$

This identity also shows why e^a can never be 0, for 0 has no reciprocal.

There is an alternate notation used for the exponential function; we write $e^x = \exp(x)$. This notation allows us to keep the exponent on the same level as the function; that can make some expressions more readable. For instance, compare

$$e^{(x^2 - 2 \cdot x^3)/2} = \exp\left((x^2 - 2 \cdot x^3)/2\right)$$

The exponential function is *increasing*: if $a < b$, then $e^a < e^b$. For instance, if $x > 0$, then $e^x > e^0 = 1$. Also, if $x < 0$, then $e^x < e^0 = 1$.

Each exponential function has an associated *logarithm* in the same base. The logarithm in base e is called the *natural logarithm* and denoted $\ln(x)$. Thus,

$$(1.1) \qquad\qquad e^a = b \quad \text{if and only if} \quad a = \ln(b)$$

There are two other ways to say the same thing.

$$(1.2) \quad \exp(\ln(b)) = b \quad \text{for all} \quad b > 0 \qquad \text{and} \qquad \ln(\exp(a)) = a \quad \text{for all} \quad a$$

We see that the natural logarithm and exponential function cancel each other when they are in sequence. This means that they are *function inverses* of each other. Function inverses are not algebraic inverses: $\ln \neq 1/\exp$; rather, the equations (1.1) define the meaning of the inverse.

The logarithm has properties analogous to the properties of the exponential function. We might have time to discuss one or two of these in class.

Properties of $\ln(x)$

    (1) $\ln(b)$ is defined for every positive number b

    (2) $\ln(1) = 0$

    (3) $\ln(b \cdot c) = \ln(b) + \ln(c)$ for all positive numbers b, c

    (4) $\ln(b^a) = a \cdot \ln(b)$ for every real number a and every positive number b

These properties are not as important as the properties of the exponential function, but properties of $\ln(x)$ will come up in various problems during the term.

Here is our first text problem. As we mentioned in the Introduction, they are indicated by the capital T.

Problem T1.1. Show that the function $\ln(x)$ is increasing.

Solution.   Suppose that $a < b$, and we need to show that $\ln(a) < \ln(b)$. (Both a, b are positive.) If not, then either $\ln(a) = \ln(b)$ or $\ln(a) > \ln(b)$.

If $\ln(a) = \ln(b)$, then (1.2) shows that

$$a = \exp(\ln(a)) = \exp(\ln(b)) = b$$

contrary to the assumption that $a < b$. Similarly, if $\ln(b) < \ln(a)$, then remembering that the exponential function is increasing, we have

$$b = \exp(\ln(b)) < \exp(\ln(a)) = a$$

again, contrary to $a < b$. We conclude that $\ln(a) < \ln(b)$, as claimed. ∎

Here is a typical problem that involves the inverse relationship between the exponential function and its logarithm.

Problem T1.2. Solve for x in the equation $3 \cdot \exp(2 \cdot x - 5) = 9$.

Solution.   Since the x we want is in the exponent, we will use the logarithm as in (1.1). In order to use those equations, we need to get the exponential function all by itself.

$$3 \cdot \exp(2 \cdot x - 5) = 9$$

$$\exp(2 \cdot x - 5) = 3 \qquad\qquad \text{now use (1.1)}$$

$$2 \cdot x - 5 = \ln(3)$$

Now it is a short step to x:

$$2 \cdot x - 5 = \ln(3) \quad \text{so that} \quad 2 \cdot x = \ln(3) + 5 \quad \text{which is} \quad x = \frac{1}{2} \cdot \left[\ln(3) + 5 \right]$$

∎

The answer we obtained is exact and, therefore, perfectly acceptable mathematically. However, we might prefer an answer that is *numerical* – an approximation obtained on a calculator or computer. A calculator gives

$$\ln(3) \approx 1.098612289$$

and so

$$x \approx \frac{1}{2} \cdot \left[(1.098612289) + 5\right] = 3.049306144$$

Mathematicians are very careful about the difference between exact answers and numerical answers. We insist on using the approximation sign $x \approx 3.049$ whenever a numerical answer occurs. There is no general rule about the number of decimal places used for numerical answers; we will be fairly loose about that.

Exact answers are almost always preferable to numerical answers. Sometimes we need to compare numbers to see which is larger, and for that purpose numerical answers can be better, provided they are accurate enough. We will do a fair amount of numerical work in the course, and so we will certainly get our fill of approximations.

We mention an algebraic detail in the previous problem. The expression $2 \cdot x - 5$ contains both a product and a difference. By the *precedence rules* of algebra, the product is done first; in other words

$$2 \cdot x - 5 = \left[2 \cdot x\right] - 5 \quad \textbf{NOT} \quad 2 \cdot \left[x - 5\right]$$

The expression on the far right would be $2 \cdot x - 10$. This applies to addition and division, as well; for instance,

$$3 \cdot x^2 + 4 \cdot x + 1/x = \left[3 \cdot x^2\right] + \left[4 \cdot x\right] + \frac{1}{x}$$

If you are likely to make a mistake in this type of expression, use parenthesis [or brackets] to clarify. If you are unsure of the meaning of the string of operations on any text problem or homework problem, please ask!

Problem T1.3. For which x is $e^{-x} < 0.01$?

Solution. A previous problem showed that the logarithm is increasing. Thus, (1.2) shows that

$$-x = \ln(\exp(-x)) < \ln(0.01) \quad \text{so that} \quad -x < \ln(0.01)$$

Multiplying both sides of the last inequality by -1, and remembering to switch the inequality, we get

$$x > -\ln(0.01) \approx 4.6052$$

■

Here are two more abstract problems. First, we claimed that *all* exponential functions can be written using the special exponential function e^x. Proving the following identity furnishes a nice exercise in using the properties of e^x.

Problem T1.4. Let $a > 0$, and let b be an arbitrary number. Show that

(1.3) $$a^b = \exp(b \cdot \ln(a))$$

Solution.   We use property (4) of the logarithm, followed by (1.2).

$$\exp(b \cdot \ln(a)) = \exp(\ln(a^b)) = a^b$$

■

For instance, $2^x = \exp(x \cdot \ln(2))$ and $10^x = \exp(x \cdot \ln(10))$.

We can also relate an arbitrary logarithm to the natural logarithm $\ln(x)$.

Problem T1.5. Let $a > 0$ and let b be an arbitrary number. Show that

$$\log_a(b) = \frac{\ln(b)}{\ln(a)}$$

Solution.   The definition of the logarithm: if $a^c = b$, then $c = \log_a(b)$. So, we prove the identity by showing that

$$a^{\ln(b)/\ln(a)} = b$$

To do this, we use the previous problem:

$$a^{\ln(b)/\ln(a)} = \exp\left[\frac{\ln(b)}{\ln(a)} \cdot \ln(a)\right] = \exp(\ln(b)) = b$$

■

2. Exponential Models

In applications of mathematics, the word *model* is used very generally for any mathematical object that represents something else. In this section we introduce some models that use the exponential function. These models are fairly simple; they introduce some aspects of models in general and they provide the opportunity to begin working with the exponential function.

Exponential Population Growth A population P of bacteria at time t grows exponentially if $P = P_0 \cdot \exp(k \cdot t)$ for some positive constant k. The number P_0 is the *initial number* of bacteria – the number of bacteria when $t = 0$.

Problem T1.6. If we start with 100 bacteria, growing exponentially, and we have 5000 in an hour, how long does it take for the population to double? Show that the doubling time does not depend on P_0 but only on k.

Solution. We will use the model to get equations to solve. We will omit some algebraic steps that use the properties (1)-(4) given above for the exponential function; you should see if you can do them yourself, and we will discuss the algebra in class.

If P is the population of bacteria after t hours, then we have

$$P = P_0 \cdot \exp(k \cdot t)$$

for some constant k. The phrase "If we start with 100" tells us that $P_0 = 100$. When $t = 1$ hour we have

$$5000 = 100 \cdot \exp(k \cdot 1) \quad \text{so that} \quad k \approx 3.912$$

To calculate doubling time and to show that it depends only on k, we think about the equation

$$2 \cdot P_0 = P_0 \cdot \exp(k \cdot t)$$

We have a specific value of k, and we can solve for t. First, divide by P_0. (What if $P_0 = 0$?) We get $2 = \exp(k \cdot t)$. This leads to

$$t = \frac{\ln(2)}{k} \approx 0.177 \text{ hours}$$

Notice that this value of t does not depend on P_0 but only on k. ■

The exact value of k in the preceding problem is $k = \ln(50)$. Remember to use the approximation sign to distinguish numerical answers! What is the exact doubling time?

Another issue may have occurred to some readers. Typical values of $P_0 \cdot \exp(kt)$ in the previous model are fractional; you might ask, "Don't population values have to be whole numbers?" Good question. Remember that a model *represents* something applied; the model is not the population! In some contexts, fractional numbers may furnish decent approximations to whole numbers; in some contexts fractional numbers are not allowed at all. We are focusing more on the arithmetic of the exponential function at this point.

Radioactive Decay   Radioactive substances lose mass over time. If Y measures the mass at time t, then we can model $Y = Y_0 \cdot \exp(-k \cdot t)$ where k is a positive constant.

Think about the following problem; we will discuss it in class.

Problem T1.7. Radium-226 is naturally radioactive. The constant k in the model is approximately $4.3 \cdot 10^{-3}$, and t is time in years. How long does it take for a mass of radium-226 to decay and lose 90% of its original mass?

Newton's Law of Heating and Cooling   We imagine placing an object in an environment of constant temperature. We assume that the object does not have any internal source of heat or cooling but is gradually warmed up or cooled off by the environment. If F is the difference between the temperature of the object and the temperature of the environment,[2] then Newton's Law of

---

[2]The temperature of the environment is also called the *ambient* temperature.

Heating and Cooling says that $F = F_0 \cdot \exp(-k \cdot t)$, where t is time and k is a positive constant, and F_0 is the initial difference in temperature.

Problem T1.8. Suppose that $F = 30$ at some value of t and $F = 20$ after 40 minutes have passed. When will $F = 0.01$?

Solution. We use 30 as our initial value of F, so that $F_0 = 30$ and then $F = 30 \cdot \exp(-k \cdot t)$. We have $F = 20$ when $t = 40$ minutes, and so

$$20 = 30 \cdot \exp(-40 \cdot k)$$

This equation can be solved for k; we get $k \approx 0.01$. We want to find t so that $F = 0.01$. We'll see that $t \approx 801$ minutes. ∎

In the problem we just did, the fact that we start with $F = 30$, a positive number, tells us that the object is warmer than the environment, and so it is cooling off. An object colder than the environment would have a negative value of F.

Turning Off a Circuit   We imagine a very simple electrical circuit consisting of a *resistor* and *capacitor* in series.[3] A circuit has a *current* I associated with it; the current is the number of electrons passing through the circuit per unit time, measured in *amperes*, usually abbreviated *amps*. In general, I is a function of time t. When our simple circuit is turned off, we have[4]

$$I = I_0 \cdot \exp\left(-\frac{t}{R \cdot C}\right)$$

where R, C are positive constants associated with the resistor and capacitor, respectively.

---

[3]It is not necessary for you to understand anything technical about electrical circuits here.

[4]It may surprise you that the current doesn't just jump to 0. According to this model, the electrons gradually slow down.

Problem T1.9. Suppose we know that $R = 10^6$ ohms, that $I_0 = 1$ amp, and that $I = 0.01$ amps when $t = 0.05$ seconds. What is C? (The units of C are *Farads*.)

Solution. As before, we substitute what we know into the equation we are given.

$$0.01 = 1 \cdot \exp\left(-\frac{0.05}{10^6 \cdot C}\right) \quad \text{so then} \quad \ln(0.01) = -\frac{0.05}{10^6 \cdot C}$$

We then get

$$C = -\frac{0.05}{10^6 \cdot \ln(0.01)} \approx 1.086 \cdot 10^{-8} \text{ Farads}$$

■

A vocabulary term: constants involved in a model are called *parameters*; the k in $\exp(k \cdot t)$, the initial values P_0, Y_0, etc., the R and C in the circuit problem, all these are parameters. Much experimental science has as its object to measure parameters.

The Cobb-Douglas Model

In the 1920's Cobb and Douglas[5] developed an equation meant to explain the production level Y of a single good as a function of capital K (monetary value) and labor L (in person-hours):

$$Y = a \cdot K^b \cdot L^c$$

where a, b, c are positive constants.[6] One of the main ways to deal with this equation is to take the logarithm of both sides, using the identities on p.3.

$$\ln(Y) = \ln(a) + b \cdot \ln(K) + c \cdot \ln(L)$$

---

[5]Cobb, C. W.; Douglas, P. H. (1928). "A Theory of Production". American Economic Review 18 (Supplement): 139–165. See also [**10**].

[6]Technically: a is the *total factor productivity*, and b, c are *output elasticities* for capital and labor, respectively.

Problem T1.10. We have the following values for a Cobb-Douglas model, where $a = 1$. Find b, c.

K	L	Y
100	25	57
80	30	54

Solution. We use the logarithmic equation. Since $a = 1$, its logarithm is 0.

$$\ln(Y) = b \cdot \ln(K) + c \cdot \ln(L)$$

$$\ln(57) = b \cdot \ln(100) + c \cdot \ln(25)$$

$$\ln(54) = b \cdot \ln(80) + c \cdot \ln(30)$$

Working with numerical values, we get

$$4.04 \approx 4.61 \cdot b + 3.22 \cdot c$$

$$3.99 \approx 4.38 \cdot b + 3.40 \cdot c$$

Solving for b, c, we obtain $b \approx 0.566$ and $c \approx 0.445$. ■

Problem T1.11. In a Cobb-Douglas model, when the capital doubles, the production increases by 23%. Find the parameter b.

Solution. We can use the original equation $Y = a \cdot K^b \cdot L^c$. For capital to double, we replace K by $2 \cdot K$. Production increases by 23%; that means that $0.23 \cdot Y$ is added to Y, resulting in

$$Y + 0.23 \cdot Y = 1.23 \cdot Y$$

Thus, we have

$$1.23 \cdot Y = a \cdot (2 \cdot K)^b \cdot L^c = 2^b \cdot a \cdot K^b \cdot L^c = 2^b \cdot Y$$

We see that $1.23 = 2^b$. Taking the logarithm of both sides, $\ln(1.23) = b \cdot \ln(2)$, and

$$b = \frac{\ln(1.23)}{\ln(2)} \approx 0.299$$

■

A Limited Population

In Chapter 2 we will meet with the *logistic equation* that describes a population growing by reproduction, but limited by an external factor. Here is a generic formula for such a population P where t is time and E, c are positive constants.

$$(1.4) \qquad P = \frac{E}{1 - c \cdot \exp(-E \cdot t)}$$

Problem T1.12. Show that P gets close to E as t gets larger and larger.[7]

Solution. We know that $E > 0$, and so $E \cdot t > 0$ when $t > 0$. As t gets larger, the quantity $-E \cdot t$ is large and negative. The graph of the exponential function shows that $\exp(-E \cdot t)$ gets smaller and smaller as t gets larger. As the exponential term gradually vanishes, the fromula (1.4) for P goes to E. ■

---

[7]This fact shows the role of the constant E in this model. It is an *equilibrium* – studied in the next chapter.

The Exponential Function.

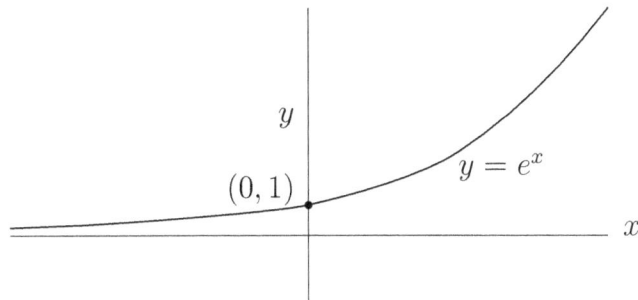

The Natural Logarithm.

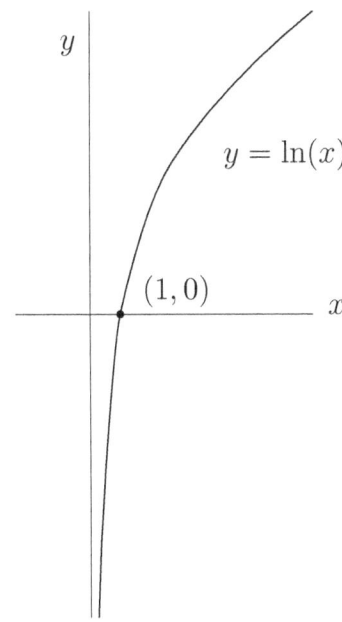

CHAPTER 2

Recursion

Suppose we monitor the population of a bacteria culture each hour, starting at 8:00 a.m. on a certain day. Let P_0 be the population at 8:00 a.m., let P_1 be the population at 9:00 a.m., let P_2 be the population at 10:00 a.m., and so on. The sequence that results

$$P_0, \; P_1, \; P_2, \; P_3, \; \cdots$$

is a model of the bacteria population. Many applications involve sequences of numbers in which a given term in the sequence determines the next term. In our bacteria population, it makes sense that the population at one time *determines* the population at the next time. Abstractly: P_n determines P_{n+1}, for each n. To be specific, let's suppose that the population increases by a factor of r each hour.[1] Thus,

$$P_{n+1} = P_n + r \cdot P_n$$

Adding $r \cdot P_n$ accomplishes the *increase*. We can write this equation like this.

$$(2.1) \qquad\qquad P_{n+1} = (1+r) \cdot P_n \quad \text{for} \quad n \geq 0$$

We have much to say about equation (2.1); we begin by calling it a *recursive equation*. It is a *recursive model* of the population. The constant r is a parameter.

---

[1]This simple model is called the *Malthusian model*, and it is closely related to the exponential model introduced in the previous chapter. See [**4**, p.2]. The entire first chapter of [**6**] is devoted to models of population growth, beginning with the Malthusian model.

In a recursive model, there is a sequence of numbers, each one determined by the ones before it. If we are given a starting number, then the rest of the sequence unfolds from there. It is important to see that equation (2.1) is really *infinitely many* equations in one. Indeed, the expression "for $n \geq 0$" tells us that the equation holds for $n = 0, 1, 2, 3, \ldots$. The number r stays the same hour to hour. Thus, (2.1) says that

$$P_1 = (1 + r) \cdot P_0 \qquad \text{(2.1) with } n = 0$$

$$P_2 = (1 + r) \cdot P_1 \qquad \text{(2.1) with } n = 1$$

$$P_3 = (1 + r) \cdot P_2$$

$$P_4 = (1 + r) \cdot P_3$$

$$\vdots$$

These equations don't tell us directly how to get the individual values, but they tell us how to *get from one value to the next*. That's recursion.

Next we observe that if we have a specific value of r and if we know the initial population P_0, then (2.1) gives us the rest of the sequence of populations. For example, let $r = 1$, so that (2.1) tells us that $P_{n+1} = 2 \cdot P_n$. Suppose that $P_0 = 3$, and just use the recursive equation over and over, as we did above:

$$P_1 = 2 \cdot P_0 = 2 \cdot 3 = 6$$

$$P_2 = 2 \cdot P_1 = 2 \cdot 6 = 12$$

$$P_3 = 2 \cdot P_2 = 2 \cdot 12 = 24$$

$$P_4 = 2 \cdot P_3 = 2 \cdot 24 = 48$$

$$\vdots$$

We can take this as far as we wish.

Given that we're doubling each time, you might know or have seen a formula for P_n in general; the point we are making is that we can produce the sequence from (2.1) without a formula. There are many recursive sequences for which no direct formula is available. And, as we mentioned above, recursion is the natural way many sequences are given in the first place. So, we want to be familiar and comfortable with recursion!

In the next section we give practical examples.

1. Recursive Models

We begin with an example that looks *mathematically* the same as Malthusian growth.

Simple Interest We have an amount of money M invested, so that every so often, *interest* is added in – we say *compounded*.[2] The interest is some fraction r of the money we have at that point in time. Suppose, for example, that interest is added each month. Let M_n be the amount of money we have after n months. Then to get M_{n+1}, we add $r \cdot M_n$.

$$(2.2) \qquad M_{n+1} = M_n + r \cdot M_n = (1 + r) \cdot M_n$$

This parameter r is the *interest rate*, and it is usually expressed as a percentage.[3] If we are given a specific interest rate and an initial amount of money M_0, then we get the sequence $M_1, M_2, \ldots$ exactly as we got the population sequence in the Malthusian model.

Repayment of a Loan We borrow an amount of money M_0 and pay it back gradually, incurring a *finance charge* as well. The finance charge is a fraction of what is owed that is added to the debt. In almost all loans, the money is paid back monthly, and so the natural time step is one month. Let

---

[2]See [**10**, Section 10.1] for an approach to interest that emphasizes a formula for M_n rather than the recursive equation.

[3]No mystery here. *Percent* means *divided by 100*, so $7\% = 0.07$. Interest rates are usually expressed as *annual* rates, and they need to be adjusted for the actual time step. If the annual rate is 6% and the time step is one month, then $r = 0.06/12 = 0.005$, since there are 12 months in a year.

r stand for the finance charge, and let q be the monthly payment, so that r, q are parameters. Then

$$(2.3) \qquad\qquad M_{n+1} = (1 + r) \cdot M_n - q$$

The finance charge is added in, and then the monthly payment is applied to decrease the amount owed. If we know the finance charge r, the monthly payment q, and the initial amount M_0, we should be able to construct the sequence $M_1, M_2, \ldots$. We are probably interested in getting M to be 0, so that the loan is payed off.

The Logistic Equation   On p.12 we introduced the *logistic equation*[4] using the exponential function. That model imagines time moving along continuously – any real number is a valid time in the model. Now we will introduce a *discrete* version of the model – a version in which time moves along step by step. For instance, we might think of time in days and make one measurement each day.

In 1838, Verhulst suggested the logistic equation for a population P that changes over time due to two phenomena: reproductive growth at a rate k and an external (environmental) limitation E. The numbers k, E are positive constants – they are parameters. Here is the discrete version of the logistic equation.

$$(2.4) \qquad\qquad P_{n+1} = P_n + k \cdot P_n \cdot \left(1 - \frac{P_n}{E}\right)$$

The units of E are those of population. You can see that this recursive equation is more complicated. Let's notice a couple of features.

An *equilibrium* is a sequence that repeats the same value over and over. We would have an equilibrium in the logistic equation if the sequence

$$P_0, P_1, P_2, P_3, \ldots \quad \text{looks like} \quad P, P, P, P, \ldots$$

---

[4]Sometimes the logistic equation is called *Verhulst's equation*; it goes by other names as well, having been rediscovered several times over the years. See [**4**, p.12]

Problem T2.1. Find the equilibria for sequences produced by the logistic equation.

Solution. We replace P_n and P_{n+1} in the logistic equation by P – that's saying that all the P_n are the same number. When we start this, we don't know, in advance, that there are any equilibria, but if there are, the equation with the P will find them. Here is that equation, and some algebra:

$$P = P + k \cdot P \cdot \left(1 - \frac{P}{E}\right)$$

$$0 = k \cdot P \cdot \left(1 - \frac{P}{E}\right)$$

Since $k > 0$, we see that

$$P = 0 \quad \text{or} \quad 1 - \frac{P}{E} = 0$$

The second equation says that $P = E$.

Remember that the equation $P = 0$ indicates the *sequence* $0, 0, 0, \ldots$; it is the sequence that is the equilibrium. We found two equilibria for the logistic equation: the sequence $0, 0, 0, \ldots$ and the sequence $E, E, E, \ldots$. We can also write $P_n = 0$ for the first sequence, remembering that $n = 0, 1, 2, \ldots$, and we write $P_n = E$ for the second sequence. The first equilibrium makes sense: if you start with no organisms, you certainly can't get any by reproduction! We will discuss the second equilibrium in class. ■

The general behavior of (2.4) is complicated and hard to deduce directly from the recursion. We will construct specific sequences for this equation to get a sense of the possibilities. The equation on p.12 hints that the exponential function is relevant. That's true but we won't be in a position to see that until much later in the course. For now, work with the discrete equation exclusively.

Speaking of equilibria, does the loan repayment model have any equilibria? To find out, write M for M_n and M_{n+1} in (2.3), and solve for M. Why is the answer to that question interesting?

Having introduced equilibria, you might ask whether there was an equilibrium in the simple interest model. It is an important question regarding just about any recursive model.

A Predator/prey model This time we have two population sequences: a population P_n of prey and a population Q_n of predators.[5] We want to keep track of both populations and their interaction, while keeping things relatively simple. The sequences P_n and Q_n will be parallel; we'll be thinking of them like this:

$$P_0 \quad P_1 \quad P_2 \quad P_3 \quad \ldots$$
$$Q_0 \quad Q_1 \quad Q_2 \quad Q_3 \quad \ldots$$

We assume there are positive constants (parameters) a, b, c, d such that

$$P_{n+1} = (1 + a) \cdot P_n - b \cdot P_n \cdot Q_n$$

$$Q_{n+1} = (1 - c) \cdot Q_n + d \cdot P_n \cdot Q_n$$

Let's give a brief explanation of these equations; we will say more about them in class. Look at the equation for P_{n+1}. The $(1+a) \cdot P_n$ term looks like Malthusian growth – reproductive growth. If $Q_n = 0$ (if there are no predators), the prey will increase in numbers exponentially. Apparently they have an abundant food supply. The $-b \cdot P_n \cdot Q_n$ term in P_{n+1} says that the prey will tend to decrease the more predators there are. The product term $P_n \cdot Q_n$ imagines random meetings between prey and predator; the greater are the numbers P_n and Q_n, the more likely these species are to encounter each other. Bad for prey.

Now look at the equation for Q_{n+1}; the $(1 - c) \cdot Q_n$ term says that the population of predators will *decrease* if left to itself (if $P_n = 0$, so that there are no prey). The term $d \cdot P_n \cdot Q_n$ represents a tendency for the predator population to increase in the presence of their food supply.

---

[5]See [**4**, p.86ff].

To get this model started, we need the initial number P_0 of prey and the initial number Q_0 of predators. Then the equations take over, producing P_1 and Q_1; then producing P_2 and Q_2, and so on.

This model has two equilibria; one of them has $P_n = 0 = Q_n$, for all n. That makes perfect sense; if you have no prey and no predators, then you will never get any! You should find the other equilibrium. Can you explain that equilibrium in ecological terms?

Let's show that the exponential model and the recursive Malthusian model are related.

Problem T2.2. In the exponential population model introduced on p.7:

$$P = P_0 \cdot \exp(k \cdot t)$$

let t measure years. Suppose that $P_0 \neq 0$. Consider the sequence $P_n = P_0 \cdot \exp(k \cdot n)$; that's the population after n years. Show that P_n follows the Malthusian model.

Solution. We need to show that the recursive equation of the Malthusian model holds: that $P_{n+1} = (1 + r) \cdot P_n$, for some number r. In other words, the ratio P_{n+1}/P_n should be a constant (the constant $1 + r$). Let's look at the ratio:

$$\frac{P_{n+1}}{P_n} = \frac{P_0 \cdot \exp(k \cdot (n+1))}{P_0 \cdot \exp(k \cdot n)} = \frac{\exp(k \cdot (n+1))}{\exp(kn)}$$
$$= \frac{\exp(kn) \cdot \exp(k)}{\exp(kn)} = \exp(k)$$

We see that $r = \exp(k) - 1$, and the sequence satisfies the Malthusian model. ∎

When k is small, $\exp(k) - 1 \approx k$, and so k and r are sometimes used interchangeably. We emphasize that this only makes sense when k (or r) is small. You might experiment on a calculator to see this.

2. Investigating Recursive Models

What might we want to know about a recursive sequence? We will use both exact formulas and numerical investigation. Regarding the latter, you will be asked to use Excel[6] on a regular basis to compute various quantities. The Spreadsheet Appendix contains lab exercises meant to introduce you to Excel; we will discuss this in class.

We begin with a very simple question about the loan repayment model.

Problem T2.3. In the loan repayment model, suppose that $10000 is borrowed, that the (annual) interest rate is 5%, and that the monthly payment is $500. How long does it take to pay off the loan?

Solution. As above, let M_n be the amount of money left to pay off after n months. Then $M_0 = 10000$. We have the recursion $M_{n+1} = (1 + r) \cdot M_n - q$. Since 5% is the annual interest rate, the monthly interest rate is $r = 0.05/12$. The payment $q = 500$.

We computed M_n for $n = 1, 2, \ldots$, looking for the place where M_n becomes negative. Here are the first few terms; the decimal parts are cents – they have been rounded.

n :	0	1	2	3	4	$\ldots$
M_n :	10000	9541.67	9081.42	8619.26	8155.18	$\ldots$

As we computed the M_n, we encountered these entries.

n :	19	20	21
M_n :	957.26	461.25	-36.83

After 20 months, there is less owed than the monthly payment, even after the finance charge is applied. So, in the 21st month, we pay off the loan. ■

Here is a new model; it measures the speed of a falling object.

---

[6]Throughout this book, *Excel* stands for *Microsoft Excel*.

Problem T2.4. We record the speed of a falling stone at each second:

$$V_0, \quad V_1, \quad V_2, \quad \ldots$$

and we suppose that $V_{n+1} = (1-k) \cdot V_n + g$, where k, g are positive parameters.[7] The units of speed here are feet per second. Find an equilibrium. Now let $k = 0.1$ and $g = 20$; starting with $V_0 = 0$, does the sequence approach the equilibrium?

Solution. To find an equilibrium, we write V for V_n and V_{n+1} and obtain the equation

(2.5) $$V = (1 - k) \cdot V + g$$

We solve $V = g/k$ feet per second. The sequence that repeats g/k over and over is an equilibrium. (The ratio g/k is called the *terminal speed* of the falling object.)

Now let $k = 0.1$ and $g = 20$, so that the equilibrium is $20/0.1 = 200$ feet per second. Letting $V_0 = 0$, we used Excel to calculate the sequence. Here are some values we obtained.

n :	0	1	2	$\cdots$	70	71	72
V_n :	0	20	38	$\cdots$	199.87	199.89	199.90

The sequence increases slowly; $V_{100} \approx 199.99$. It looks like the sequence is approaching equilibrium. ∎

The Babylonian Sequence Let $c > 0$, and define a recursion

$$a_{n+1} = \frac{1}{2} \cdot \left(a_n + \frac{c}{a_n} \right) \quad \text{for} \quad n = 0, 1, 2, \ldots$$

Heron of Alexandra[8] attributed this formula to ancient Babylonian mathematicians. To find out what it is for, we experiment, starting with $a_0 = c$.

---

[7]The number k measures the air resistance on the stone; the number g measures the pull of gravity.

[8]See [**14**, p.423].

When $c = 1$, compute that the sequence is in equilibrium $1, 1, 1, \ldots$. When $c = 4$, here are numerical values of the first few terms.

$$4, \ , 2.5, \ 2.05, \ 2.00061, \ 2.00000009$$

It looks like the sequence is approaching the number 2. Is that an equilibrium when $c = 4$? We take $c = 4$ and $a_0 = 2$, then we get $2, 2, 2, \ldots$.

Problem T2.5. Show that $\sqrt{c}$ is an equlibrium for the Babylonian Sequence.

Solution.   We let $a_n = \sqrt{c}$, and show that $a_{n+1} = \sqrt{c}$:

$$a_{n+1} = \frac{1}{2} \cdot \left(\sqrt{c} + \frac{c}{\sqrt{c}} \right) = \frac{1}{2} \cdot \left(\sqrt{c} + \frac{\sqrt{c}^2}{\sqrt{c}} \right)$$

$$= \frac{1}{2} \cdot \left(\sqrt{c} + \sqrt{c} \right) = \frac{1}{2} \cdot 2 \cdot \sqrt{c} = \sqrt{c}$$

∎

The Babylonian Sequence is used to approximate square roots, and it gets a good deal of accuracy very quickly. You should try approximating $\sqrt{2}$ for yourself: use $c = 2 = a_0$.

A Markov Process   We begin with a specific, fanciful example. Suppose that we have exactly two types of weather: sunny and rainy. If today is sunny, it is 70% likely to be sunny tomorrow and 30% likely to be rainy. If today is rainy, it is 55% likely to be sunny tomorrow and 45% likely to be rainy.

Suppose that on day n, the probability of sunny is S_n, and the probability of rainy is R_n. We want to calculate the probability of sun and rain for the next day. The probabilities will be denoted S_{n+1} and R_{n+1}. To calculate S_{n+1}, we consider the possible weather the day before (day n). First, it could have been sunny and then it went from sunny to sunny. The probability of sun on day n was S_n and the probability of sun on day n and sun on the next day is 70% (from the previous paragraph). The elementary theory of probability tells us that the probability of the chain of events is a product, so that the probability of being sunny on day n and then going to sunny on day $n+1$ is $S_n \cdot 0.7$.

The other way we could be sunny on day $n+1$ is to be rainy on day n and then go from rain to sun: $R_n \cdot 0.55$. Since these are the two ways we get sunny on day $n+1$, we have

$$S_{n+1} = 0.7 \cdot S_n + 0.55 \cdot R_n$$

We add the probabilities here, since the two events (sunny day n, rainy day n) do not overlap.

We calculate R_{n+1} in a similar way. See if you can do it, and we'll discuss it in class.

$$R_{n+1} = 0.3 \cdot S_n + 0.45 \cdot R_n$$

As in the predator/prey model, we have two recursive sequences that are related to each other. You will experiment with the behavior of these sequences.

A Markov process involves a sequence of time steps; at each step there are a finite number of possibilities, exactly one of which must happen at that step. The possibilities are usually called *states*. The states in the weather example are sunny and rainy. For each states S, T, there is a definite probability associated with having state S at time step t and state T at time step $t+1$. We will see several examples of Markov processes.

The Fibonacci Numbers In 1202, Fibonacci[9] became interested in an idealized problem involving populations of rabbits. Here is what he assumed.

(1) we always have many available males to mate
(2) immature female rabbits take a month to become mature enough to reproduce

---

[9]This problem is a typical counting problem that involves recursion. Books on *combinatorics*, such as [**16**] contain many such problems. See [**15**, p.214ff] for information on Fibonnaci.

(3) mature female rabbits gestate each month, become pregnant, and produce one male and one female at the end of the month

Fibonacci was able to find a recursion for the number F_n of female rabbits at the end of month n:

$$(2.6) \qquad F_{n+2} = F_{n+1} + F_n \quad \text{for} \quad n = 0, 1, 2, \ldots$$

This is a different type of recursion than what we have considered so far. In the previous examples, you computed a sequence term from the term just previous to it. In (2.6) you need *two* previous terms.

Can you show that (2.6) is correct? You might want to consider the number M_n of mature females at the end of month n, and the number I_n is immature females. Then $F_n = M_n + I_n$. Fibonacci's assumptions show that

$$M_{n+1} = M_n + I_n$$
$$I_{n+1} = M_n$$

We will complete the details in class.

If you assume you have one immature female to start with, so that $M_0 = 0$ and $I_0 = 1$, then you get $F_0 = 1$ and $F_1 = 1$, and then the recursion (2.6) takes over. This specific example gives what are called the *Fibonacci numbers*; they have a remarkable number of interesting properties and they arise in a host of counting problems.

We have tried to keep to fairly simple models in this chapter, while pushing in the direction of the complexity found in more realistic applications. The reader might glance at Chapter 6 of [6], regarding gas exchange in the human lung, to see a typical, much more complicated, example.

CHAPTER 3

The Derivative

This chapter begins our analytic work.

1. Discovering the Derivative

Calculus is about variation; it attempts to relate the change in one variable to the change in others. Here is a standard notation for change: if the variable x starts at the value 3 and ends up at the value 5, then we write $\Delta x = 5 - 3 = 2$ to indicate the *change in x*. The capital delta Δ stands for *change*. In general, change is computed as

$$\Delta = (\text{final value}) - (\text{starting value})$$

The expression Δx is not a product; it is one symbol (a word spelled with two letters). The value of Δx depends on two values of x; one value designated as starting and other as final.

To work another example, suppose the variable z starts at 7 and ends up at 3. Then

$$\Delta z = 3 - 7 = -4$$

Note that the change is *final minus start* regardless of the size of either number. In this case, Δz was negative – that means that the variable z *decreased*. In the first example $\Delta x = 2$ was positive, indicating that x *increased*.

Notice that if x is the starting value and the change is Δx, then the ending value is $x + \Delta x$.

Later we will consider the situation of an object moving along a line – say that the line is the x-axis. The value of x at some point in time is the object's *position*. The quantity Δx, the change in position, is called the *displacement*. If we undergo a displacement of -2, we mean that we move 2 units in the negative direction on the x-axis; that would be to the left.

The delta notation will arise frequently with two variables related by a function. Say $y = f(x)$. Starting with a value of x, we have the corresponding value of $y = f(x)$. Now say we undergo a change Δx in x, the ending value of x is $x + \Delta x$, and the y-value that goes with this is $f(x + \Delta x)$. So, the starting value of y is $f(x)$ and the ending value of y is $f(x + \Delta x)$, and

$$(3.1) \qquad \Delta y = f(x + \Delta x) - f(x)$$

This equation is important; it relates the change in a variable y to values of a function $f(x)$ that computes that variable y.

Now we move toward this chapter's main topic. Let's consider two application problems: finding the velocity of a moving object and finding the slope of a graph. We will see that these problems are, in some sense, the same problem; when we abstract out what the two problems have in common, we will discover the *derivative*. This is very typical mathematics – a general idea is discovered as a common feature of several concrete problems. An additional application will be discussed in class.

Problem: Velocity.

We are interested in the discovery of a mathematical idea; it is interesting that the approach of a particular physics text is identical. See [**9**, p.23ff].

We begin with some definitions that may be familiar to you – we want to make sure everyone is on the same page. We imagine traveling along a straight line – let's say we are traveling along the x-axis. The *speed* of a moving body measures the ratio of distance traveled to the time elapsed. A constant speed of 30 miles per hour (MPH) predicts that we travel 30 miles each hour elapsed.

The *velocity* of a moving body is the speed along with a plus or minus sign that indicates the direction of the motion: positive velocity means that we are traveling in the positive direction on the x-axis (to the right); negative velocity means that we are traveling to the left. So, if the velocity is -10 MPH, then the speed is 10 MPH and we are moving to the left. Notice that speed and velocity have the same units.

If the speed is constant, then we can calculate it by dividing distance traveled by time elapsed. Say we start at $x = 3$ at time $t = 5$, and we end up at $x = -10$ when $t = 7$. Then we have traveled a distance of $3 + 10 = 13$ miles in 2 hours, and so our speed was $13/2 = 6.5$ MPH. The velocity, on the other hand, would be the *displacement* divided by the time elapsed, for displacement takes direction into account. The displacement in this example is

$$\Delta x = (\text{ending } x) - (\text{starting } x) = (-10) - (3) = -13$$

and so the velocity is $-13/2 = -6.5$ MPH. Notice that the denominator 2 in the case of both speed and velocity is the *change in time* Δt. In particular, notice that the ratio x/t is not relevant to either quantity: for instance, we had $x = 3$ when $t = 7$; the ratio $3/7$ doesn't tell us about speed.

Constant velocity is very unusual. For instance, motion often starts from a standstill, speeding up and slowing down along the way. It is the business of calculus to explain how to calculate velocity when it is *not constant*. Let's think about that.

Suppose that the x axis measures miles, and time t is in hours. Suppose we want to know the velocity when $t = 3$. Let's assume we are at $x = 10$ when $t = 3$, and we end up at $x = -80$ at $t = 5$. The displacement is $\Delta x = -80 - 10 = -90$ miles, and the elapsed time is $\Delta t = 5 - 3 = 2$ hours. Their ratio is

$$\frac{\Delta x}{\Delta t} = \frac{-90}{2} = -45 \ \text{ miles per hour}$$

What does the number -45 tell us? It would be too much to expect that it tells us the velocity at $t = 3$, for the velocity may have varied quite a bit during the two hours elapsed. The ratio gives what is called an *average velocity*. An average is meant to give one number that smooths out variation along with way, but it doesn't tell us much about what was happening exactly at $t = 3$.

Instead of measuring over an elapsed time of 2 hours, suppose we use $\Delta t = 0.25$ hours; say we start at $t = 3$ and end at $t = 3.25$, and suppose that the displacement is $\Delta x = -12$ miles over this change in time. Then the average velocity is

$$\frac{\Delta x}{\Delta t} = \frac{-12 \text{ miles}}{0.25 \text{ hours}} = -48 \text{ miles per hour}$$

The time interval 0.25 hours is much smaller than 2 hours, and so we expect that -48 miles per hour is closer to the actual velocity at $t = 3$. Nonetheless, there could be *some* variation in 0.25 hours, and so -48 miles per hour is still an *average velocity*, and we still don't know our *exact velocity*.

Now suppose we have $\Delta t = 0.01$ hours, and the displacement is $\Delta x = -0.47$ miles. This time, the average velocity is

$$\frac{\Delta x}{\Delta t} = \frac{-0.47 \text{ miles}}{0.01 \text{ hours}} = -47 \text{ miles per hour}$$

Because we are measuring over one hundredth of an hour we expect even less variation in velocity than with the previous 0.25 hours, so, this average velocity is probably closer to the actual velocity at $t = 3$.

It looks like this: to get the average velocity to be close to the actual velocity, use a very small elapsed time, so that the velocity can't vary too much. As the time interval gets smaller and smaller, we expect the average velocity to be close to the actual velocity. The average velocity is

$$\frac{\Delta x}{\Delta t}$$

and here is the technical notation for letting the time interval Δt get very small.

$$(3.2) \qquad \text{velocity} = \lim_{\Delta t \to 0} \frac{\Delta x}{\Delta t}$$

The notation $\Delta t \to 0$ is meant to convey that Δt (the elapsed time) gets smaller and smaller; we consider the average velocity ratio for each possible value of Δt, and see what happens. The notation lim stands for *limit*, and we read $\lim_{\Delta t \to 0}$ as "the limit as Δt goes to 0." Limits can be studied for their own sake; in this course they will occur in a fairly restricted set of instances.

Here is a very specific example. Suppose that $x = t^2$, and say we want the velocity when $t = 3$. The position at $t = 3$ is $x = 3^2 = 9$. The equation (3.2) asks for Δx and Δt. If t starts at 3 and undergoes a change Δt, then it ends up at $3 + \Delta t$, and x then ends up at $(3 + \Delta t)^2$. Thus,

$$\Delta x = \text{ending} - \text{starting} = (3 + \Delta t)^2 - 9$$

We indulge in some algebra on the average velocity. At the start, we use the standard formula

$$(3.3) \qquad (a + b)^2 = a^2 + 2 \cdot a \cdot b + b^2$$

Here goes.

$$\begin{aligned}
\frac{\Delta x}{\Delta t} &= \frac{(3 + \Delta t)^2 - 9}{\Delta t} = \frac{9 + 6 \cdot \Delta t + (\Delta t)^2 - 9}{\Delta t} \\
&= \frac{6 \cdot \Delta t + (\Delta t)^2}{\Delta t} = \frac{\Delta t \cdot (6 + \Delta t)}{\Delta t} \qquad \text{cancel } \Delta t \\
&= 6 + \Delta t
\end{aligned}$$

We see that the average velocity is actually quite simple in this case. When Δt is very close to 0, the expression $6 + \Delta t$ is close to 6. This shows that the *exact velocity* at $t = 3$ is 6.

Problem: Slope

We are given a curve $y = f(x)$ in the plane, and we consider the problem of finding the slope at some point. The slope is, by definition, the slope of a *tangent line* to the curve. If we are given a sketch of the curve, we can probably determine a tangent line by eye; what we want is an exact, analytic way of finding the slope of that line. Idea: we can find the slope of a line through two points on the curve – such a line is a *secant line*. If the two points are very close together, their secant line is approximately a tangent line.

Suppose we are interested in the slope at some point (x_0, y_0) on the curve $y = f(x)$. We will imagine a secant line through (x_0, y_0) and some other point on the curve. We can imagine this other point as resulting from a change of Δx in x, and then the *ratio* $\Delta y/\Delta x$ gives the slope of the secant line.

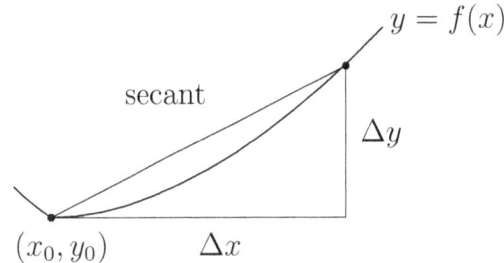

If the two points at the ends of the secant are very close to each other – if Δx is small – it makes sense that the secant line would be close to the tangent. In other words,

$$\text{tangent slope} = \lim_{\Delta x \to 0} \frac{\Delta y}{\Delta x}$$

Let's work a specific example. Let's find the slope to $y = 1 - x^2$ at $x = 2$. The point on the curve where $x = 2$ is $(2, -3)$. That's our starting point for secant lines. Stepping Δx, we have x-coordinate $2 + \Delta x$, and then the y-coordinate is $1 - (2 + \Delta x)^2$. And so, again using (3.3), we compute

$$\Delta y = 1 - (2 + \Delta x)^2 - (-3) = 1 - \left[4 + 4 \cdot \Delta x + (\Delta x)^2\right] + 3$$
$$= 1 - 4 - 4 \cdot \Delta x - (\Delta x)^2 + 3 = -4 \cdot \Delta x - (\Delta x)^2$$

Then the slope of the secant line is

$$\text{secant slope} = \frac{\Delta y}{\Delta x} = \frac{-4 \cdot \Delta x - (\Delta x)^2}{\Delta x} = \frac{\Delta x \cdot (-4 - \Delta x)}{\Delta x} = -4 - \Delta x$$

When Δx is small, this expression is close to -4. The slope of the tangent to $y = 1 - x^2$ at the point $(2, -3)$ is -4.

There are many other examples that can be given; each one constitutes an application of the derivative. Here is one that we will discuss in class.

Problem: Density

Assume we have a block of wood lying along the interval $0 \leq x \leq 2$ on the x-axis, where the x measures yards. (So the wood is 2 yards long.) We want to think about the weight of the wood. Notice that it doesn't make sense to talk about the weight at a point along the block, since a point is not three-dimensional, and so the weight at a point is 0 (ideally). We measure how weight *accumulates* as we move from left to right over the log. For each x with $0 \leq x \leq 2$, let $W(x)$ be the weight in pounds of the wood from the left-hand end up to the point x. In other words, $W(x)$ is the weight of x yards of wood – the left-hand x yards. We will see that this function $W(x)$ can be used to calculate the *density* of the wood – the number of pounds per yard we have at any given point.[1]

---

[1]The density can vary over the block, either because the thickness of the block varies, or because the characteristics of the wood vary. This approach to density – using the function W that measures the weight at various points is consistent with the ideas of Isaac Newton that led to the discovery of the general calculus. See [**15**]. Also, we mention that the word *density* means different things in different contexts; we will point this out as we encounter examples.

At a point $x = 1$ on the block, we contemplate a change Δx. We think of Δx as giving us a small part of the block. Here is a picture, imagining that $\Delta x > 0$.

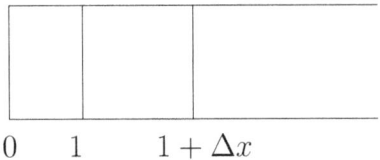

$$0 \quad 1 \qquad 1 + \Delta x$$

By definition $W(1)$ is the weight of the block up to the point $x = 1$, and $W(1 + \Delta x)$ is the weight up to $x = 1 + \Delta x$. Thus,

$$\Delta W = W(1 + \Delta x) - W(1)$$

is the weight in pounds of the block *between* $x = 1$ and $x = 1 + \Delta x$. The ratio

$$\frac{\Delta W}{\Delta x}$$

has units pounds per yard, an *approximation* of the density at x. Why approximation? Because the thickness or shape could vary between 1 and $1 + \Delta x$. In fact, the ratio $\Delta W / \Delta x$ is the *average density* on that piece of wood. Letting Δx get smaller and smaller, we expect to get closer to the actual density at x.

You might try the following problem.

Problem T3.1. Let $W(x) = 2 + x^3$, and calculate the density at $x = 1$.

2. The Derivative at a Point

In our calculation of velocity the independent variable was time t, whereas in the calculation of slope the independent variable was x. However, the calculations were similar in that they involved a ratio of deltas. In both cases, we computed something like this:

(3.4)
$$\lim_{\Delta x \to 0} \frac{\Delta y}{\Delta x}$$

where the steps Δx were taken from a specific point (x_0, y_0) on the curve $y = f(x)$, so that $y_0 = f(x_0)$. The limit (3.4) is called the *derivative of $f(x)$ at $x = x_0$.*

Problem T3.2. Compute the derivative of $1/x$ at $x = 4$.

Solution. At the point $x = 4$ on the curve $y = 1/x$, we have $y = 1/4$. Thus, x starts at 4 and y starts at $1/4$. The definition (3.4) of the derivative wants Δy:

$$\Delta y = \frac{1}{4 + \Delta x} - \frac{1}{4} = \frac{4 - (4 + \Delta x)}{(4 + \Delta x) \cdot 4} = \frac{-\Delta x}{4 \cdot (4 + \Delta x)}$$

And so

$$\frac{\Delta y}{\Delta x} = \frac{-\Delta x}{4 \cdot (4 + \Delta x)} \cdot \frac{1}{\Delta x} = \frac{-1}{4 \cdot (4 + \Delta x)}$$

We see that when Δx is small, the ratio is close to $-1/(4 \cdot 4)$. Thus, the slope of $y = 1/x$ at $x = 4$, is $-1/16$. Or, if $y = 1/x$ describes an object traveling on the y-axis, where x is time, then we just computed the velocity of the object at time $x = 4$. ■

Problem T3.3. Compute the derivative of $3 \cdot x - 7$ at $x = -2$.

Solution. On $y = 3 \cdot x - 7$ and $x = -2$, we have $y = -13$. Then check that

$$\Delta y = 3 \cdot (-2 + \Delta x) - 7 - (-13) = 3 \cdot \Delta x$$

and so

$$\frac{\Delta y}{\Delta x} = \frac{3 \cdot \Delta x}{\Delta x} = 3$$

The resulting expression 3 stays at that value no matter what Δx does. Thus, the derivative is 3. ■

Let's interpret the derivative we just calculated. That derivative is the *slope* of $y = 3 \cdot x - 7$; that curve is a straight line of slope 3, so it's no wonder that the derivative is 3.

There are many other examples. We may have time to mention some of them in class.

(1) Let $v(t)$ be the velocity at time t of an object moving along an axis. Then the derivative of v at $t = t_0$ is the *acceleration* $a(t)$ of the object at time $t = t_0$. The units of acceleration are distance/(time)^2.

(2) If $W(x)$ is the work done in moving an object along the x-axis to the point x, then the derivative of W at $x = x_0$ is the force applied at the point $x = x_0$ to produce that motion.

(3) If $C(x)$ measures the cost of quantity x of some good, then the derivative of C at $x = x_0$ is the cost per unit of x, at $x = x_0$. (The cost per unit is the *price*; price can vary, depending on how much we buy.)

You might wonder about calculating the derivative *numerically*.

Problem T3.4. Approximate the derivative of e^x at $x = 2$.

Solution. Let $y = e^x$. To compute Δy, we use 2 as our starting value and then $2 + \Delta x$ will be our ending value. Then

$$\Delta y = \exp(2 + \Delta x) - \exp(2) \quad \text{and} \quad \frac{\Delta y}{\Delta x} = \frac{\exp(2 + \Delta x) - \exp(2)}{\Delta x}$$

We used Excel to compute the ratio for various small values of Δx:

x:	0.1	0.05	0.01	0.005	0.001	0.0001	0.00001
$\Delta y / \Delta x$:	7.7711	7.5769	7.4261	7.4076	7.3928	7.3894	7.3891

It looks as if the derivative is 7.389, or so. ∎

In the next chapter we will be able to find the exact value of the derivative of e^x; numerical calculation raises the question why we shouldn't prefer that method to the algebraic method we used to this point. There are two important reasons to prefer the algebraic method: (1) in applications of the derivative we will want to understand the behavior of the derivative over a large set of points, all at once. Algebraic formulas disclose that behavior much more readily than tables of values. (2) Many of the (practical!) uses of the derivative involve purely symbolic quantities – there will be no specific numerical values to compute. Having said that, we do admit that numerical calculations are often useful, and they are necessary in situations where we do not have a formula for the function involved. We close by saying that the numerical calculation of derivatives, in general, is a hard problem with many nuances. We won't take the time to introduce that subject in any formal way. Still, there are a couple of problems at the end of this chapter that invite numerical calculation.

3. The Derivative of a Function

We have been calculating the derivative at a specific point. Given a function such as x^2, it turns out to be efficient to calculate the derivative *at all points simultaneously.* To do that, we imagine the curve $y = x^2$ at a generic point (x, x^2). We start at x and move to $x + \Delta x$. We can compute Δy:

$$\Delta y = (x + \Delta x)^2 - x^2$$

Expanding out:

$$\Delta y = x^2 + 2 \cdot x \cdot \Delta x + (\Delta x)^2 - x^2$$
$$= 2 \cdot x \cdot \Delta x + (\Delta x)^2$$

And so

$$\frac{\Delta y}{\Delta x} = \frac{2 \cdot x \cdot \Delta x + (\Delta x)^2}{\Delta x} = 2 \cdot x + \Delta x$$

In the expression $2 \cdot x + \Delta x$, the x indicates s a specific point on the curve – we just don't want to make *which* point. And the Δx is supposed to get small. When that happens, the expression $2 \cdot x + \Delta x$ gets close to $2 \cdot x$ – that's the value of the derivative of x^2, in general.

Problem T3.5. Find the slope of $y = x^2$ at $x = 5$.

Solution. The slope of the tangent line is *the derivative of x^2 at $x = 5$.* We calculated the derivative $2 \cdot x$ *for all x,* and so the derivative at $x = 5$ is simply $2 \cdot 5 = 10$. Thus, 10 is the slope of the tangent line at $x = 5$. ■

Here is one of the notations we will use to designate the general derivative:

$$(x^2)' = 2 \cdot x$$

In this notation, the prime mark indicates the derivative.

Let's see what the derivative looks like in general. Say we have a function $f(x)$. If $y = f(x)$, and if we step from x to $x + \Delta x$, then (3.1) (back on p.28) said that

$$\Delta y = f(x + \Delta x) - f(x)$$

And so,

$$\frac{\Delta y}{\Delta x} = \frac{f(x + \Delta x) - f(x)}{\Delta x}$$

To get the derivative, we need to let $\Delta x \to 0$, as in (3.4). We can write the result in two ways:

$$(3.5) \qquad f'(x) = \lim_{\Delta x \to 0} \frac{f(x + \Delta x) - f(x)}{\Delta x}$$

or

$$(3.6) \qquad f'(x) = \lim_{\Delta x \to 0} \frac{\Delta y}{\Delta x}$$

Problem T3.6. Find $(1/x)'$.

Solution. With $f(x) = 1/x$, we imitate (3.5). To do this, it might help to write $f(x)$ with a placeholder instead of the x.

$$f(\Box) = \frac{1}{\Box} \quad \text{so that} \quad f(x + \Delta x) = \frac{1}{x + \Delta x}$$

And (3.5) looks like this.

$$\frac{f(x + \Delta x) - f(x)}{\Delta x} = \frac{1/(x + \Delta x) - 1/x}{\Delta x}$$

To work with this, we first write the denominator as its own fraction:

$$\frac{1/(x + \Delta x) - 1/x}{\Delta x} = \left[\frac{1}{x + \Delta x} - \frac{1}{x} \right] \cdot \frac{1}{\Delta x}$$

Now we continue.

$$\frac{f(x + \Delta x) - f(x)}{\Delta x} = \left[\frac{1}{x + \Delta x} - \frac{1}{x}\right] \cdot \frac{1}{\Delta x} \quad \text{get a common denominator}$$

$$= \frac{x - (x + \Delta x)}{(x + \Delta x) \cdot x} \cdot \frac{1}{\Delta x}$$

$$= \frac{-\Delta x}{(x + \Delta x) \cdot x} \cdot \frac{1}{\Delta x} \quad \text{cancel } \Delta x$$

$$= \frac{-1}{(x + \Delta x) \cdot x}$$

Let's not lose track of this expression; it's nothing more nor less than $\Delta y / \Delta x$. As Δx gets close to 0, the expression looks like $-1/x^2$.

$$\left(\frac{1}{x}\right)' = -\frac{1}{x^2}$$

■

Here is a case where we can predict the answer before we compute it.

Problem T3.7. Compute the derivative of $m \cdot x + b$, where m, b are constants.

Solution. This time we will use (3.6). Let $y = m \cdot x + b$ is a line of slope m. Since the derivative computes slope, we expect that

$$(m \cdot x + b)' = m$$

Let's check. We have

$$\Delta y = \left[m \cdot (x + \Delta x) + b\right] - \left[m \cdot x + b\right]$$

$$= m \cdot x + m \cdot \Delta x + b - m \cdot x - b$$

$$= m \cdot \Delta x$$

And then

$$\frac{\Delta y}{\Delta x} = \frac{m \cdot \Delta x}{\Delta x} = m$$

This secant slope is constant. (That's because a *secant line* to a line is just the line itself!) And so the derivative is m. ■

We have seen that similar patterns can occur in variables with different names. If we have $f(B)$, then $f'(B)$ asks for the derivative of f, just as $f'(x)$ does. Thus, $(B^2)' = 2 \cdot B$, just as we already computed $(x^2)' = 2x$. Once we compute the derivative of a function abstractly, we have the derivative in the same function no matter what variable is used. That efficiency is the point of having general formulas.

Suppose we have $y = f(x)$ and we know that $f'(3) > 0$. We claim that y is *increasing* when x increases from $x = 3$. This is because $\Delta y/\Delta x \to f'(3)$ as $\Delta x \to 0$, with $x = 3$. Since $f'(3)$ is positive, the ratios $\Delta y/\Delta x$ must be positive when Δx is small. If x increases from 3 a very small bit, then $\Delta x > 0$, and so for $\Delta y/\Delta x$ to be positive, we must have Δy positive, as well. That Δy is positive is that y is increasing.

A similar argument can be made when $f'(x) < 0$. In that case, we say that y is *decreasing*, which means that if x increases, then y decreases. This gives very practical information that we will use often.

Virtually every function for which there is a formula has a derivative that we can calculate, as we will see. But there are some (unusual) functions for which the ratio in (3.5) does not do anything sensible as Δx goes to 0. The functions $f(x)$ for which $f'(x)$ exists are said to be *differentiable*. The most common way for us to know that a function is differentiable is to have a formula for its derivative; in the next few sections of this text we will develop a set of rules for taking the derivative – rules which will apply to most every function formula.

When we have $y = f(x)$, the equations (3.5) and (3.6) write the derivative two ways:

$$(3.7) \qquad f'(x) = \lim_{\Delta x \to 0} \frac{f(x + \Delta x) - f(x)}{\Delta x} = \lim_{\Delta x \to 0} \frac{\Delta y}{\Delta x}$$

These formulations are really the same, since

$$\Delta y = f(x + \Delta x) - f(x)$$

The formulations represent different points of view. The ratio that uses $f(x)$ is thinking about the derivative as applying to a function; the ratio with Δy is thinking about the relationship between the variables x, y.

These two points of view are united in one more notation. It is customary to write

$$(3.8) \qquad \frac{dy}{dx} = \lim_{\Delta x \to 0} \frac{\Delta y}{\Delta x}$$

Equation (3.6) shows that dy/dx is just another notation for the derivative:

$$(3.9) \qquad \frac{dy}{dx} = f'(x)$$

Why have two notations for the same thing? As we mentioned, the two notations represent two different points of view. It turns out that the dy/dx notation will be useful in its own right. The dy and dx are *differentials*.[2] Similar to the Δx notation, dx is one symbol and not a product. It is not trivial to explain exactly what differentials are; for our course, we will see that they furnish a useful notation, especially when units are involved. Indeed, suppose that x measures weight in pounds and y measures frogs, and suppose there is a function $y = f(x)$ that tells us how many frogs we get for so-many pounds. (We can suppose that different frogs have different weights, so $f(x)$ may be quite complicated.) The units of the ratio $\Delta y/\Delta x$ would be frogs per pound, that ratio coming directly from the units of y and the units of x. Since $\Delta y/\Delta x \to dy/dx$, the derivative dy/dx has the same units: frogs per pound.

This simple reasoning works in general. If we have an object moving along the x-axis, where that axis measures feet, and if t is time in seconds, then the

---

[2]The differential notation dy/dx comes from Leibnitz, one of the discoverers of the calculus. The book [15] has more information.

velocity $v(t)$ has units feet per second. The derivative formula

$$v(t) = \frac{dx}{dt}$$

shows the same ratio of units: feet per second. Taking this one step further, the derivative of $v(t)$ could be written

$$v'(t) = \frac{dv}{dt} \quad \text{in units} \quad \frac{\text{feet per second}}{\text{second}}$$

and a little fraction algebra:

$$\frac{\text{feet per second}}{\text{second}} = \frac{\text{feet}}{\text{second}} \cdot \frac{1}{\text{second}} = \frac{\text{feet}}{\text{second}^2}$$

The units feet per second-squared are the units of *acceleration*, and in fact the derivative of the velocity is acceleration, as we have mentioned.

If $W = f(x)$ is the weight in pounds of the left-hand x feet of a steel bar, then

$$f'(x) = \frac{dW}{dx} \quad \text{pounds/foot}$$

As we know from our previous work, dW/dx is the *density*.

CHAPTER 4

Computing the Derivative

Let's summarize the notation for the derivative. Given $y = f(x)$, we have two ways to write the derivative:

$$(4.1) \qquad f'(x) = \frac{dy}{dx}$$

The definition of the derivative asks us to think about a ratio; here are the two formulations of that ratio.

$$(4.2) \qquad \frac{\Delta y}{\Delta x} = \frac{f(x + \Delta x) - f(x)}{\Delta x}$$

The ratio of deltas is the *average* change in y per change in x. When we let $\Delta x \to 0$, we get the derivative, the *exact* change in y per change in x:

$$(4.3) \qquad f'(x) = \lim_{\Delta x \to 0} \frac{f(x + \Delta x) - f(x)}{\Delta x}$$

A function that has a derivative is a *differentiable function*. In terms of differentials:

$$(4.4) \qquad \frac{dy}{dx} = \lim_{\Delta x \to 0} \frac{\Delta y}{\Delta x}$$

The units of $f'(x)$ are the units of y divided by the units of x.

We are about to learn rules for calculating the derivative of a function from the form of the function, without using the limits (4.3) and (4.4) directly. We have already learned one such rule: the derivative of $m \cdot x + b$ is the slope m.

We remind you of a few other special cases that occur often.

$$c' = 0 \quad \text{for all constants} \quad c$$

$$(x^2)' = 2x$$

$$\left(\frac{1}{x}\right)' = -\frac{1}{x^2}$$

In this chapter we will lay out the standard derivative formulas learned by everyone taking calculus at every level. It doesn't take a lot of verbiage to write out all the formulas; to learn them you it will help to work as many problems as possible for each formula. The problems document has exercises of all types!

1. The Power Rule, Linearity

Power Rule Let n be a number. Then the derivative of x^n is $n \cdot x^{n-1}$. In other words

$$\text{If} \quad y = x^n \quad \text{then} \quad \frac{dy}{dx} = (x^n)' = n \cdot x^{n-1}$$

For instance

$$(x^2)' = 2 \cdot x \quad \text{and} \quad (x^3)' = 3 \cdot x^2 \quad \text{and} \quad (x^4)' = 4 \cdot x^3 \quad \text{and so on}$$

Another example: since $1/x = x^{-1}$, the Power Rule applies here, as well, for that rule would compute

$$\left[x^{-1}\right]' = (-1) \cdot x^{-2} = -\frac{1}{x^2}$$

and this was what we got by the limit calculation of the derivative. We will prefer writing the derivative of $1/x$ as $-1/x^2$.

We will discuss a general argument for the Power Rule later. Here are some special cases, using the limit (3.5). We will skirt some of the details, in case you want to try them yourself!

Problem T4.1. Show that $(x^3)' = 3 \cdot x^2$.

Solution. If $y = x^3$, then[1]

$$\Delta y = (x + \Delta x)^3 - x^3 = 3 \cdot x^2 \cdot \Delta x + 3 \cdot x \cdot (\Delta x)^2 + (\Delta x)^3$$

and so

$$\frac{\Delta y}{\Delta x} = 3 \cdot x^2 + 3 \cdot x \cdot \Delta x + (\Delta x)^2$$

As $\Delta x \to 0$, we get $(x^3)' = 3 \cdot x^2$. ∎

Problem T4.2. Show that $(x^{-2})' = -2 \cdot x^{-3}$.

Solution. Let $y = 1/x^2$, and then

$$\Delta y = \frac{1}{(x + \Delta x)^2} - \frac{1}{x^2}$$
$$= \frac{x^2 - (x + \Delta x)^2}{(x + \Delta x)^2 \cdot x^2}$$
$$= \frac{-2 \cdot x \cdot \Delta x - (\Delta x)^2}{(x + \Delta x)^2 \cdot x^2}$$

and so

$$\frac{\Delta y}{\Delta x} = \frac{-2 \cdot x - \Delta x}{(x + \Delta x)^2 \cdot x^2}$$

As $\Delta x \to 0$, we get

$$\frac{dy}{dx} = \frac{-2 \cdot x}{x^4} = \frac{-2}{x^3} = -2 \cdot x^{-3}$$

∎

Here are two more rules will allow us to calculate the derivative of a wide variety of functions.

Constant Multiple Rule If k is a constant, and $f(x)$ is differentiable, then

$$\left(k \cdot f(x)\right)' = k \cdot f'(x)$$

Problem T4.3. Compute $(4 \cdot x^2)'$.

---

[1]We are using an expansion of $(x + \Delta x)^3$ that is analogous to the expansion of $(a + b)^2$ in (3.3). These are special cases of what is called the *Binomial Theorem*.

Solution. Using the Constant Multiple Rule, we have $\left(4 \cdot x^2\right)' = 4 \cdot \left(x^2\right)'$. We know that $(x^2)' = 2 \cdot x$, and so

$$\left(4 \cdot x^2\right)' = 4 \cdot (x^2)' = 4 \cdot 2 \cdot x = 8 \cdot x$$

∎

Another case that comes up often: a minus sign is multiplication by -1, so that

$$[-x^3]' = \left[(-1) \cdot x^3\right]' = (-1) \cdot [x^3]' = (-1) \cdot 3 \cdot x^2 = -3 \cdot x^2$$

When we take the derivative of a negative, we don't have to convert the minus sign to a (-1). We just pull out the minus sign; the point is that we are pulling out a constant according to the Constant Multiple Rule: $\left[-f(x)\right]' = -f'(x)$.

Addition Rule If $f(x)$ and $g(x)$ are differentiable, then

$$\left(f(x) + g(x)\right)' = f'(x) + g'(x)$$

Example. We have

$$\left(\frac{1}{x} + x^5\right)' = \left(\frac{1}{x}\right)' + (x^5)'$$

We already know the derivative of $1/x$ and of x^5, and so

$$\left(\frac{1}{x} + x^5\right)' = \left(\frac{1}{x}\right)' + (x^5)' = -\frac{1}{x^2} + 5 \cdot x^4$$

∎

The Constant Multiple Rule and the Addition Rule together are sometimes called the *linearity* of the derivative. Here is an abstract version that combines them: let $f(x)$ and $g(x)$ be differentiable functions, and let a, b be constants. Then

$$\left[a \cdot f(x) + b \cdot g(x)\right]' = \left[a \cdot f(x)\right]' + \left[b \cdot g(x)\right]' \qquad \text{Addition}$$

$$= a \cdot f'(x) + b \cdot g'(x) \qquad \text{Constant Multiple}$$

Example.   Here we combine the power rule and linearity; underneath each line we give the reason for that particular step.

$$\left(4 \cdot x^5 - 3 \cdot x^3 + 7 \cdot x + 1\right)'$$

$$= \left(4 \cdot x^5\right)' + \left(-3 \cdot x^3\right)' + \left(7 \cdot x\right)' + \left(1\right)'$$

addition rule

$$= 4 \cdot \left(x^5\right)' - 3 \cdot \left(x^3\right)' + 7 \cdot \left(x\right)' + 0$$

constant multiple, derivative of a constant

$$= 4 \cdot 5 \cdot x^4 - 3 \cdot 3 \cdot x^2 + 7 \cdot 1$$

power rule, including $x' = 1$

$$= 20 \cdot x^4 - 9 \cdot x^3 + 7$$

∎

Example.   Slightly uglier functions.

$$\left(3 \cdot x^2 + \frac{9}{x} - x^{-4}\right)' = \left(3 \cdot x^2\right)' + \left(9 \cdot \frac{1}{x}\right)' + \left(-x^{-4}\right)'$$

addition rule

$$= 3 \cdot \left(x^2\right)' + 9 \cdot \left(\frac{1}{x}\right)' + (-1) \cdot \left(x^{-4}\right)'$$

constant multiple

$$= 3 \cdot 2 \cdot x + 9 \cdot (-1) \cdot \frac{1}{x^2} + (-1) \cdot (-4) \cdot x^{-5}$$

power rule

$$= 6 \cdot x - \frac{9}{x^2} + 4 \cdot x^{-5}$$

algebra

∎

In the last two examples we belabored the steps to make sure you could see exactly what we were doing. As we do more and more derivative calculations, we will work at doing them more quickly, using the rules without thinking about them too much. In getting started, it will help to be self-conscious of the rules we are using.

Problem T4.4. We are moving along the y-axis; position on that axis is measured in feet. At time t seconds we are at $y = 3t - 12t^3$. What is our velocity when $t = -2$?

Solution.   Velocity is dy/dt feet per second. The derivative:

$$\frac{dy}{dt} = \left(3t - 12t^3\right)' = 3 \cdot 1 - 12 \cdot 3 \cdot t^2 = 3 - 36 \cdot t^2$$

That's the velocity at an arbitrary time t. When $t = -2$, we get velocity

$$3 - 36 \cdot (-2)^2 = 3 - 36 \cdot 4 = -141 \text{ feet per second}$$

■

Problem T4.5. An object travels along the y-axis (measuring meters in its usual vertical orientation) such that $y = 3 \cdot t^2 - 12 \cdot t + 14$ at time t seconds. When is the object moving down?

Solution.   "Moving down" means that the velocity dy/dt is negative. We compute

$$\frac{dy}{dt} = \left(3 \cdot t^2 - 12 \cdot t + 14\right)' = 3 \cdot 2 \cdot t - 12 = 6t - 12 \text{ meters per second}$$

We want the velocity to be negative:

$$6t - 12 < 0 \quad \text{which is} \quad 6t < 12 \quad \text{that is} \quad t < 2$$

Before $t = 2$, the object is moving down. ■

If Q is a quantity that depends on time, then $Q' = dQ/dt$ measures the *rate of change* of Q. Say, for example, that Q measures kg, and t is time in seconds. Then Q' has units kg/sec. Suppose that $Q = 10$ kg and $Q' = 2$ kg/sec. It is common to express the rate of change 2 as a percentage of the value of Q. The number 2 is 2/10=20% of the number 10. The ratio 20% is the *relative rate* of Q. Thus, the relative rate of Q is

$$\frac{Q'}{Q} = \frac{1}{Q} \cdot \frac{dQ}{dt}$$

Of course, we must be assuming that $Q \neq 0$. Notice the units of relative rate here: percent per second.

Many uses of the derivative involve abstract quantities for which we do not have equations. Here is an example.

Problem T4.6. Helium gas in a closed container obeys the *Ideal Gas Law:*[2] $V \cdot P = k \cdot T$, where P is the pressure, V the volume, T the absolute temperature, and k a positive constant. In a closed container, we would have V constant, whereas P, T could change in time. Show that P, T have the same relative rates.

Solution.   We take the derivative of both sides of the Gas Law equation, using time t as variable. Since V, k are constants, the Constant Multiple Rule applies to them:

$$V \cdot \frac{dP}{dt} = k \cdot \frac{dT}{dt}$$

To get at relative rates, we divide each side of this equation by the corresponding side of the original equation:

$$\frac{1}{VP} \cdot V \cdot \frac{dP}{dt} = \frac{1}{kT} \cdot k \cdot \frac{dT}{dt} \quad \text{so that} \quad \frac{1}{P} \cdot \frac{dP}{dt} = \frac{1}{T} \cdot \frac{dT}{dt}$$

This says that P, T have the same relative rate. ∎

Since P, T have the same relative rates, if P is changing by, say, 23% per second, then T is changing at 23% per second, as well.

---

[2]See [**9**, p.543].

2. Products and Quotients

We introduce two more fundamental rules for calculating derivatives. We will focus on understanding and using these rules. In class we will consider the question of their validity.

Product Rule Given differentiable functions $f(x)$ and $g(x)$, we have

$$\big(f(x) \cdot g(x)\big)' = f'(x) \cdot g(x) + f(x) \cdot g'(x)$$

This rule is sometimes described as *taking turns*; we take the derivative of the first function, leaving the second alone, and then we leave the first function alone and take the derivative of the second function.

Example. We imitate the Product Rule carefully, and then apply the other rules previously discussed. Make sure you can account for this calculation!

$$\left[(x^2 - 4 \cdot x + 3) \cdot (5 - 1/x)\right]'$$
$$= (x^2 - 4 \cdot x + 3)' \cdot (5 - 1/x) + (x^2 - 4 \cdot x + 3) \cdot (5 - 1/x)'$$
$$= (2 \cdot x - 4) \cdot (5 - 1/x) + (x^2 - 4 \cdot x + 3) \cdot (0 - (-1)/x^2)$$
$$= (2 \cdot x - 4) \cdot (5 - 1/x) + (x^2 - 4 \cdot x + 3) \cdot (1/x^2)$$

The expression that results is somewhat complicated; nonetheless, we do not try to simplify it, since we are just interested in applying the rules. ■

The rules of differentiation work together. Here is a calculation using the Product Rule – but it could be done another way.

$$(x^3)' = (x \cdot x^2)' = (x)' \cdot x^2 + x \cdot (x^2)'$$
$$= 1 \cdot x^2 + x \cdot 2 \cdot x = x^2 + 2 \cdot x^2 = 3 \cdot x^2$$

The answer agrees with the Power Rule; in fact, the Product Rule *implies* the Power Rule in the case that the exponent is a positive integer.

Here is a practical, abstract use of the Product Rule.

Problem T4.7. Suppose we sell tables, and the weekly *demand* D for tables (how many we can sell) depends on the unit price p. The notation $D(p)$ expresses that demand is a function of price. Suppose that we don't know $D(p)$ exactly, but we know that it is positive and we know that $D'(p) < 0$. Under what circumstances should we raise the price in order to increase revenue?

Solution. That $D'(p) < 0$ is that $D(p)$ *decreases* when p *increases.*[3]

The *revenue* R is the product of demand and price: $R = D(p) \cdot p$; this is the product of the number of tables and the price of each table. To have an increase in price result in an increase in revenue, we want $dR/dp > 0$.

The Product Rule computes

$$\frac{dR}{dp} = D'(p) \cdot p + D(p) \cdot 1$$

In the inequality

$$D'(p) \cdot p + D(p) > 0$$

we want to divide by $D'(p)$. Since that quantity is negative, the division will *reverse* the inequality:

$$p + \frac{D(p)}{D'(p)} < 0 \quad \text{which is} \quad p < -\frac{D(p)}{D'(p)}$$

Again we remind you that $D'(p)$ is negative, and so the fraction $-D(p)/D'(p)$ is positive. The last inequality says that if the price is below the ratio $-D/D'$, then we should raise the price to increase R (even though we will decrease demand). ■

Problem T4.8. Use the Product Rule to show that $(f(x)^2)' = 2 \cdot f(x) \cdot f'(x)$.

Solution. Write $f(x)^2 = f(x) \cdot f(x)$, and use the Product Rule:

$$\left[f(x) \cdot f(x) \right]' = f'(x) \cdot f(x) + f(x) \cdot f'(x) = 2 \cdot f(x) \cdot f'(x)$$

■

---

[3]This fact about the sign of the derivative was mentioned on p.40.

The derivative rule for ratios of functions is somewhat similar to the Product Rule.

Quotient Rule   Given differentiable functions $f(x)$ and $g(x)$, we have

$$\left(\frac{f(x)}{g(x)}\right)' = \frac{f'(x)\cdot g(x) - f(x)\cdot g'(x)}{g(x)^2}$$

The numerator of this formula should remind you of the Product Rule: we *take turns* with each function. But there is a minus sign in between rather than a plus sign. We need to take the derivative of the *numerator* $f(x)$ first.

Example.   The Quotient Rule used carefully.

$$\left[\frac{3\cdot x - 4}{x^5 + 7\cdot x + 2}\right]'$$

$$= \frac{(3\cdot x - 4)'\cdot (x^5 + 7\cdot x + 2) - (3\cdot x - 4)\cdot (x^5 + 7\cdot x + 2)'}{(x^5 + 7\cdot x + 2)^2}$$

$$= \frac{3\cdot (x^5 + 7\cdot x + 2) - (3\cdot x - 4)\cdot (5\cdot x^4 + 7)}{(x^5 + 7\cdot x + 2)^2}$$

As in our first Product Rule example, there is no need to work further with this expression if we are just trying to apply the rule. ■

As with the Product Rule, the Quotient Rule is consistent with the other rules.

Example.   We calculate the derivative of x^{-5} using the Quotient Rule, applying the Power Rule to x^5.

$$(x^{-5})' = \left(\frac{1}{x^5}\right)' \qquad\qquad \text{algebra}$$

$$= \frac{1'\cdot x^5 - 1\cdot (x^5)'}{(x^5)^2} \qquad\qquad \text{Quotient Rule}$$

$$= \frac{0 \cdot x^5 - 5 \cdot x^4}{x^{10}} \qquad\qquad \text{Power Rule } x^5$$

$$= \frac{-5 \cdot x^4}{x^{10}} \qquad\qquad\qquad \text{algebra}$$

$$= -5 \cdot x^{4-10} = -5 \cdot x^{-6}$$

The final answer agrees with the Power Rule applied to x^{-5}. ■

We mentioned that the Product Rule can be used to establish the Power Rule for x^n when n is a positive integer. The previous example can be generalized to establish the Power Rule when n is a negative integer, using the Quotient Rule. Thus, the Product Rule and Quotient Rule give the Power Rule when the exponent is an integer.

Problem T4.9. The mass M kg of a cluster of cells satisfies $M = t/(t^2 + 9)$, where t is time in hours. For $t \geq 0$, when is the mass increasing?

Solution. Recall that M is increasing when $dM/dt > 0$. (By the way, the derivative has units kg per hour.) We compute dM/dt using the Quotient Rule.

$$\frac{dM}{dt} = \frac{1 \cdot (t^2 + 9) - t \cdot 2 \cdot t}{(t^2 + 9)^2}$$

We need to work with this derivative to find out where it is positive. So, we simplify the numerator:

$$\frac{1 \cdot (t^2 + 9) - t \cdot 2 \cdot t}{(t^2 + 9)^2} = \frac{t^2 + 9 - 2 \cdot t^2}{(t^2 + 9)^2} = \frac{-t^2 + 9}{(t^2 + 9)^2}$$

We want to know where this is positive. The denominator $(t^2 + 9)^2$ is positive, and so the fraction is positive when the numerator is positive. That's

$$-t^2 + 9 > 0 \quad \text{which is} \quad 9 > t^2$$

Since $t \geq 0$, we see that $9 > t^2$ means that $3 > t$. Thus, M is increasing when $t < 3$. ■

3. **Exponentials and Logarithms**

In Chapter 1 we introduced exponential functions A^x and the function called *the* exponential function e^x. Also, we have been enjoying the alternative notation $e^x = \exp(x)$. Be sure you remember the graph of $y = e^x$ and of the logarithm $y = \ln(x)$; those graphs are on p.13. The prominence of e^x comes from the simplicity of its derivative.

The Exponential Function We have

$$\left(e^x\right)' = e^x \quad \text{and} \quad \exp'(x) = \exp(x)$$

This is a remarkable formula – it takes some experience to appreciate it. Let's focus on using it, at first.

Problem T4.10. For which x does $y = \exp(x)/(x+3)$ have positive slope?

Solution. We are asking when $dy/dx > 0$. Use the Quotient Rule:

$$\left[\frac{e^x}{x+3}\right]' = \frac{(e^x)' \cdot (x+3) - e^x \cdot (x+3)'}{(x+3)^2} = \frac{e^x(x+3) - e^x}{(x+3)^2} = \frac{e^x(x+2)}{(x+3)^2}$$

(Notice, in the second equation, when we took the derivative of the e^x, it doesn't look as if anything happened! That's the simplicity of the derivative rule for the exponential function.) We want our slope to be positive. The exponential function is positive, and the $(x+3)^2$ in the denominator is positive, and so the sign of the slope is determined by $x + 2$. We see that the slope is positive when $x > -2$. ∎

The graph of the exponential function shows that it is always increasing. In other words, its slope is always positive – but that slope is just the exponential function, which is positive, so that makes sense.

On p.36, we estimated the derivative of e^x at $x = 2$ numerically. Now we see that the exact value of that derivative is e^2. The formula $(e^x)' = e^x$ depends on the limit considered in the following.

Problem T4.11. Show that $(\exp(\Delta x) - 1)/\Delta x \approx 1$ when Δx is small. Then assume that $(\exp(\Delta x) - 1)/\Delta x \to 1$ as $\Delta x \to 0$, and show that the derivative formula for $\exp(x)$ follows.

Solution. Excel gives the following approximations:

$$
\begin{array}{cccccc}
\Delta x: & 0.1 & 0.01 & 0.001 & 0.0001 & 0.00001 \\
(e^{\Delta x} - 1)/\Delta x: & 1.051709 & 1.00502 & 1.000500 & 1.000050 & 1.000005
\end{array}
$$

Not definitive but seemingly consistent with the claimed value 1.

Now assume the limit, and we use the properties of the exponential function to manipulate the delta-ratio.

$$
\begin{aligned}
\frac{\exp(x + \Delta x) - \exp(x)}{\Delta x} &= \frac{\exp(x) \cdot \exp(\Delta x) - \exp(x)}{\Delta x} \\
&= \frac{\exp(x) \cdot [\exp(\Delta x) - 1]}{\Delta x} \\
&= \exp(x) \cdot \frac{\exp(\Delta x) - 1}{\Delta x}
\end{aligned}
$$

The ratio to the right goes to 1, by assumption, and so the original fraction goes to $\exp(x)$, and that's the derivative of $\exp(x)$. ■

The limit we assumed in the previous problem requires more theoretical knowledge of the exponential function than we are willing to pursue, so we will be content to assume that the derivative formula for e^x is correct. Later in the text, we will see that this formula leads to the following.

The Natural Logarithm   We have

$$
\ln'(x) = \frac{1}{x}
$$

Problem T4.12. Where does $x \cdot \ln(x)$ have derivative equal to 0?

Solution. Product rule:

$$
\left[x \cdot \ln(x) \right]' = 1 \cdot \ln(x) + x \cdot \frac{1}{x} = \ln(x) + 1
$$

This is zero when $\ln(x) + 1 = 0$, so that $\ln(x) = -1$, and that's $x = e^{-1} = 1/e$.
∎

Problem T4.13. Explain why $\ln(4x)$ and $\ln(x)$ have the same derivative.

Solution. We need to remember an identity from Chapter 1:

$$\ln(4x) = \ln(4) + \ln(x)$$

The number $\ln(4)$ is a constant, and its derivative is 0. And so,

$$\left[\ln(4x)\right]' = \left[\ln(4) + \ln(x)\right]' = 0 + \frac{1}{x}$$

That's why $\ln(4x)$ and $\ln(x)$ have the same derivative. ∎

4. The Chain Rule

A *composite function* is formed by plugging one function into another. For instance, suppose that $f(y) = y^2 + 5 \cdot y$ and $g(x) = 3x - 2$, then

$$f(g(x)) = g(x)^2 + 5 \cdot g(x) = (3x - 2)^2 + 5 \cdot (3x - 2)$$

The *Chain Rule* explains how to take the derivative of a composite function.

Chain Rule We have

$$\left[f(g(x))\right]' = f'(g(x)) \cdot g'(x)$$

Let's go right to an example: consider

$$\left[(x^4 + 3 \cdot x^2)^5\right]'$$

The pattern $f(g(x))$ sees an outside function f and an inside function g. The outside function in our example is the 5-th power function. The inside function is $x^4 + 3 \cdot x^2$. The Chain Rule says to take the derivative of the outside function, leaving the inside function alone. That's the expression $f'(g(x))$ in the Chain

Rule formula. The derivative of the 5-th power function comes from the Power Rule; here is the formula, using $\square$ as the variable.

$$\left[\square^5\right]' = 5 \cdot \square^4$$

In our example, $\square = x^4 + 3 \cdot x^2$, so we have

$$f'(g(x)) = 5 \cdot \left(x^4 + 3 \cdot x^2\right)^4$$

The other factor in the Chain Rule formula is $g'(x)$; that's the derivative of the inside function:

$$\left[x^4 + 3 \cdot x^2\right]' = 4 \cdot x^3 + 3 \cdot 2 \cdot x = 4 \cdot x^3 + 6 \cdot x$$

Putting these together:

$$[f(g(x))]' = f'(g(x)) \cdot g'(x)$$
$$\left[(x^4 + 3 \cdot x^2)^5\right]' = 5 \cdot \left(x^4 + 3 \cdot x^2\right)^4 \cdot (4 \cdot x^3 + 6 \cdot x)$$

∎

Let's do another one; this time we will try to focus on the outside-then-inside pattern and not clutter things with function notation.

$$\left[\frac{1}{(5 - 7 \cdot x^4)^7}\right]' = \left[(5 - 7 \cdot x^4)^{-7}\right]'$$
$$= -7 \cdot (5 - 7 \cdot x^4)^{-8} \cdot (5 - 7 \cdot x^4)' \qquad \text{***}$$
$$= -7 \cdot (5 - 7 \cdot x^4)^{-8} \cdot (0 - 7 \cdot 4 \cdot x^3)$$

At the equation marked (***) we took the derivative of the (-7)th power (the outside function), leaving the inside function $5 - 7 \cdot x^4$ alone. Then we took the derivative of the inside.

Problem T4.14. Find where is y decreasing, if

$$y = \frac{3}{(x^2 + 5)^2}$$

Solution. We want to find where $dy/dx < 0$. The expression for y is a quotient, and we could use the Quotient Rule, but let's notice that

$$\frac{3}{(x^2 + 5)^2} = 3 \cdot (x^2 + 5)^{-2}$$

We read $(x^2 + 5)^{-2}$ as an outside function (the (-2) power) and an inside function $x^2 + 5$. Here is the derivative:

$$
\begin{aligned}
\left[3 \cdot (x^2 + 5)^{-2}\right]' &= 3 \cdot \left[(x^2 + 5)^{-2}\right]' & \text{constant multiple} \\
&= 3 \cdot (-2) \cdot (x^2 + 5)^{-3} \cdot (x^2 + 5)' & \text{Chain Rule} \\
&= -6 \cdot (x^2 + 5)^{-3} \cdot 2 \cdot x & \\
&= \frac{-12 \cdot x}{(x^2 + 5)^3} & \text{algebra}
\end{aligned}
$$

Where is this negative? The denominator is always positive, so we would need the numerator to be negative. We have $-12 \cdot x < 0$ when $x > 0$. ∎

Problem T4.15. Find the derivative of e^{3t}.

Solution. The function e^{3t} is a composite; this is easier to see if we use the notation $\exp(3t)$; the outside function is the exponential function and the inside function is $3t$. Thus, we use the Chain Rule:

$$\left[\exp(3t)\right]' = \exp'(3t) \cdot (3t)' = \exp(3t) \cdot 3 = 3e^{3t}$$

The derivative formula $\exp' = \exp$ was used, leaving the inside $3t$ alone. ∎

The previous problem can be done another way; write $e^{3t} = (e^t)^3$, and then

$$\left[(e^t)^3\right]' = 3 \cdot (e^t)^2 \cdot (e^t)' = 3e^{2t} \cdot e^t = 3e^{3t}$$

Same answer!

We take the previous problem a little further. Many of the exponential models in Chapter 1 were of the form $y = y_0 \cdot \exp(k \cdot t)$, where y_0, k are constant. We can explain these models as showing that y changes at a rate proportional to y. Indeed, the rate of change of y is

$$\frac{dy}{dt} = \Big[y_0 \cdot \exp(k \cdot t)\Big]' = y_0 \cdot \exp(k \cdot t) \cdot (k \cdot t)' = y_0 \cdot \exp(k \cdot t) \cdot k = k \cdot y$$

The expression $k \cdot y$ is *proportional to* y, with k being the constant of proportionality. Thus, y changes at a rate proportional to y. Furthermore, the *relative rate* of y is constant:

$$\frac{1}{y} \cdot \frac{dy}{dt} = \frac{1}{y} \cdot k \cdot y = k$$

For example, suppose we have a population that grows at a relative rate of 10% per year. Then $k = 10\% = 0.1$, and the population P satisfies the equation $P = P_0 \cdot \exp(0.1 \cdot t)$, where t is time in years.

Problem T4.16. Where is the slope to $y = \exp(-x^2)$ positive, and where is it negative?

Solution. The slope is the derivative, and the derivative uses the Chain Rule, as in the last problem.

$$y' = \exp'(-x^2) \cdot (-x^2)' = \exp(-x^2) \cdot (-2 \cdot x) = -2 \cdot x \cdot \exp(-x^2)$$

The exponential function is always positive, and so it doesn't affect the sign of y'. We see that $y' > 0$ when $-2x > 0$, and this happens when $x < 0$. Thus, the curve $y = \exp(-x^2)$ is increasing when $x < 0$. We have $y' < 0$ when $x > 0$, and this is where $y = \exp(-x^2)$ is decreasing. ∎

Problem T4.17. An object moves along the y-axis (in meters) such that $y = \ln(t^2 - t + 1)$, where t is time in seconds. When is the object moving up?

Solution. (Note: the quantity $t^2 - t + 1$ is always positive, and so the logarithm is defined.) We are moving up when the velocity dy/dt is positive.

$$\frac{dy}{dt} = \Big[\ln(t^2 - t + 1)\Big]' = \ln'(t^2 - t + 1) \cdot (t^2 - t + 1)'$$

$$= \frac{1}{t^2 - t + 1} \cdot (2t - 1) = \frac{2t - 1}{t^2 - t + 1}$$

As we mentioned, the quantity $t^2 - t + 1$ is positive, and so the sign is determined by $2t - 1$. We have $2t - 1 > 0$ when $t > 1/2$. After $1/2$ second we are moving up. ■

Problem T4.18. Show that the relative rate of $f(t)$ is the derivative of $\ln(f(t))$, assuming that $f(t)$ is positive.

Solution. Chain Rule again!

$$\big[\ln(f(t))\big]' = \ln'(f(t)) \cdot f'(t) = \frac{1}{f(t)} \cdot f'(t)$$

as needed. ■

We have made the point previously that the differentiation rules are related to each other.

Problem T4.19. Show that the Chain Rule and the Product Rule imply the Quotient Rule.

Solution. Let $f(x), g(x)$ be differentiable and write

$$\frac{f(x)}{g(x)} = f(x) \cdot g(x)^{-1}$$

To take the derivative, we apply the Product Rule and then the Chain Rule.

$$\left[\frac{f(x)}{g(x)}\right]' = \left[f(x) \cdot g(x)^{-1}\right]' \qquad\qquad \text{product}$$

$$= f'(x) \cdot g(x)^{-1} + f(x) \cdot \left[g(x)^{-1}\right]' \qquad\qquad \text{chain on } g^{-1}$$

$$= f'(x) \cdot g(x)^{-1} + f(x) \cdot (-1) \cdot g(x)^{-2} \cdot g'(x) \qquad\qquad \text{algebra}$$

$$= \frac{f'(x)}{g(x)} - \frac{f(x)}{g(x)^2} = \frac{f'(x)g(x) - f(x)g'(x)}{g(x)^2} \qquad\qquad ■$$

Why is the Chain Rule true? We will not give a complete proof, but we can present the main idea of the proof. In order to compute $[f(g(t))]'$, we need to consider the ratio

$$\frac{f(g(t + \Delta t)) - f(g(t))}{\Delta t}$$

We will rewrite the numerator, letting $x = g(t)$, so that $\Delta x = g(t + \Delta t) - g(t)$, and then

$$\frac{f(g(t + \Delta t)) - f(g(t))}{\Delta t} = \frac{f(x + \Delta x) - f(x)}{\Delta t}$$

Supposing that $\Delta x \neq 0$, we have

$$\frac{f(x + \Delta x) - f(x)}{\Delta t} = \frac{f(x + \Delta x) - f(x)}{\Delta x} \cdot \frac{\Delta x}{\Delta t}$$

We get the derivative of $f(g(t))$ if we let $\Delta t \to 0$.

Letting $\Delta t \to 0$, the ratio $\Delta x / \Delta t$ goes to $dx/dt = g'(t)$. Also, it makes sense that if $\Delta t \to 0$, then $\Delta x \to 0$, and so

$$\frac{f(x + \Delta x) - f(x)}{\Delta x} \to f'(x) = f'(g(t))$$

Putting the two factors back together, we get

$$\left[f(g(t))\right]' = \lim_{\Delta t \to 0} \frac{f(x + \Delta x) - f(x)}{\Delta x} \cdot \frac{\Delta x}{\Delta t} = f'(g(t)) \cdot g'(t)$$

and that's the Chain Rule.

The foregoing raises two technical issues; although we will not resolve either of them, honesty compels us to point them out. First, we have to consider what happens when $\Delta x = 0$. Second, we need the fact that $\Delta x \to 0$ when $\Delta t \to 0$. That second fact is the *continuity of* $x = g(t)$; you can look up continuity in Section 10-3 of [1], if you are curious.

The Chain Rule can be used to explain the Power Rule. Recall the formula (1.3) on p.6; it says that

$$x^n = \exp(n \cdot \ln(x))$$

(We need $x > 0$ so that $\ln(x)$ is defined.) Taking the derivative with respect to x:

$$(x^n)' = \exp'(n \cdot \ln(x)) \cdot (n \cdot \ln(x))' = \exp(n \cdot \ln(x)) \cdot \frac{n}{x} = x^n \cdot \frac{n}{x} = n \cdot x^{n-1}$$

and that's the Power Rule, at least when $x > 0$. We'll be content with that case.

CHAPTER 5

Interpreting and Using the Derivative

We begin with an algebraic preliminary: the problem of determining where a function is positive, where it is negative, and where it is zero. For a *rational function* (a polynomial or a ratio of polynomials), we can answer the question by factoring.

Problem T5.1. Let $f(x) = x^4 - x^3 - 6 \cdot x^2$. Find out for where $f(x)$ is positive and where it is negative.

Solution. The *roots* of $f(x)$ are the x values where $f(x)$ is 0. It is a theorem of elementary algebra that $x = a$ is a root of $f(x)$ if and only if $x - a$ is a factor of $f(x)$. Thus, to find where $f(x)$ is 0, we factor, starting with noticing the x^2 on each term. For the other factors, there are several methods you may have learned; we will be happy to help you individually if you need some review.

$$x^4 - x^3 - 6 \cdot x^2 = x^2 \cdot (x^2 - x - 6) = x^2 \cdot (x + 2) \cdot (x - 3)$$

The factors show that the roots of $f(x)$ are $x = 0, -2, 3$. We graph these on a number line, and this chops the real numbers into four intervals:

$$x < -2, \quad -2 < x < 0, \quad 0 < x < 3, \quad 3 < x$$

On each of these intervals $f(x)$ is not 0, and so on each interval its sign cannot change.[1] There is a fairly simple way to figure out these signs, working from

---

[1]The assertion that f cannot change sign *between* roots follows from the fact already enunciated that roots come from factors. There is also a calculus theorem called the Intermediate Value Theorem that is relevant to this situation. You are encouraged to look up that result online or in a calculus text.

right to left along the intervals. For the rightmost interval $3 < x$, the sign is determined by the *leading coefficient* of the polynomial – the coefficient on the highest power of x. That highest power is x^4 and the coefficient is 1; thus, the sign on $3 < x$ is *positive*. If you wish, choose various values of x greater than 3; plug each of them into the polynomial to observe that you get a positive value each time.

As we move to the left, across each root, we look at the *exponent* of that root as a factor of $f(x)$. When we cross $x = 3$, going from $3 < x$ to $0 < x < 3$, the factor is $x - 3$ to the first power. The exponent is 1 is an *odd* number, and so the sign *changes* as we cross $x = 3$. Thus, the sign of $0 < x < 3$ is *negative*.

Moving left across $x = 0$, the factor is x^2 and the exponent 2 is *even*. When we cross a root with an even exponent, the sign stays the same. The sign of $0 < x < 3$ was negative, and so the sign of $-2 < x < 0$ is negative, too.

When we cross $x = -2$, we encounter the factor $x + 2$ which has exponent 1, an odd exponent, so that the sign changes. Here is the resulting *sign diagram*.

$$+ \qquad - \qquad - \qquad +$$
$$\underline{\hspace{1cm}} \ -2 \ \underline{\hspace{1cm}} \ 0 \ \underline{\hspace{1cm}} \ 3 \ \underline{\hspace{1cm}}$$

For any number x, the diagram tells us quickly whether $f(x)$ is positive, negative, or zero. For instance $f(-1) < 0$ and $f(17) > 0$ and $f(1) < 0$. ∎

Problem T5.2. Find the signs of $(x^2 - 4 \cdot x - 5)/(x - 7)$.

Solution.   Factor!
$$\frac{x^2 - 4 \cdot x - 5}{x - 7} = \frac{(x + 1) \cdot (x - 5)}{x - 7}$$
The roots are $x = -1$ and $x = 5$. The denominator root $x = 7$ is a *singularity* – a place where the function is not defined. We include singularities in the determination of sign, in the same way as roots. The leading coefficient on the function is 1, and each exponent is 1. Here is the sign diagram.

$$- \qquad + \qquad - \qquad +$$
$$\underline{\hspace{1cm}} \ -1 \ \underline{\hspace{1cm}} \ 5 \ \underline{\hspace{1cm}} \ 7 \ \underline{\hspace{1cm}}$$

Problem T5.3. Find the signs of $10 \cdot x^4 - 5 \cdot x^3 - 5 \cdot x^5$.

Solution. We rearrange terms to go from the highest power of x to the lowest; then we see we can pull out $-5 \cdot x^3$.

$$10 \cdot x^4 - 5 \cdot x^3 - 5 \cdot x^5 = -5 \cdot x^5 + 10 \cdot x^4 - 5 \cdot x^3$$
$$= -5 \cdot x^3 (x^2 - 2 \cdot x + 1)$$
$$= -5 \cdot x^3 \cdot (x - 1)^2$$

Here is the sign diagram; make sure you can explain each of the signs!

$$\begin{array}{ccccc} + & & - & & - \\ \underline{} & 0 & \underline{} & 1 & \underline{} \end{array}$$

■

We will work other examples in class.

1. Curve Sketching

We show how to make a *quick sketch* of a curve $y = f(x)$. We will see that a great deal of information can be observed from such a graph.

We know that $dy/dx = f'(x)$ is the slope of the curve $y = f(x)$. Also, $y = f(x)$ is increasing when $f'(x) > 0$, and it is decreasing when $f'(x) < 0$. If $f'(x) = 0$ when $x = a$, then $x = a$ is called a *critical point*. At each critical point, the tangent has slope 0; the tangent is horizontal.

By a *quick sketch* we mean a sketch that plots critical points and that indicates where a curve is increasing and where is it decreasing. The increasing/decreasing feature of a curve is its *oscillation*. These features are determined by the signs of the derivative.

Problem T5.4. Draw a quick sketch of $y = x^2 - 6 \cdot x + 8$.

Solution. We find the sign diagram for the derivative, as in the last section.

$$\frac{dy}{dx} = 2 \cdot x - 6 = 2 \cdot (x - 3)$$

This expression is 0 when $x = 3$, it is positive when $x > 3$, and it is negative when $x < 3$. These are signs of the *derivative*. To relate them back to the curve: when $x > 3$ we have $y' > 0$, so that y is increasing. When $x < 3$, we have $y' < 0$, so that y is decreasing.

The point $x = 3$ where $y' = 0$ is a critical point. When $x = 3$ we have $y = -1$, so the critical point $(3, -1)$ is on the graph. Here is a sketch of the graph, using the information just obtained.

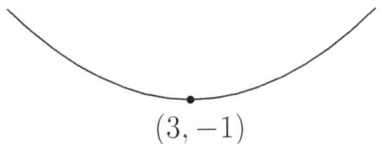

$(3, -1)$

■

Problem T5.5. Make a quick sketch of $y = (x^2 + x + 4)/(x + 1)$.

Solution. The derivative uses the Quotient Rule; the first equation below applies that rule carefully. We skip lightly over the rest of the algebra:

$$y' = \frac{(2 \cdot x + 1) \cdot (x + 1) - (x^2 + x + 4) \cdot 1}{(x + 1)^2}$$

$$= \frac{x^2 + 2 \cdot x - 3}{(x + 1)^2} = \frac{(x + 3) \cdot (x - 1)}{(x + 1)^2}$$

The critical points: $x = -3$ and $x = 1$. Here is the sign diagram, including the resulting information about the graph. Note carefully the use of y' as opposed to y. We use the abbreviations *inc* and *dec* for *increasing* and *decreasing*, respectively.

$$y': \quad + \qquad - \qquad - \qquad +$$
$$\underline{\quad\quad} \ -3 \ \underline{\quad\quad} \ -1 \ \underline{\quad\quad} \ 1 \ \underline{\quad\quad}$$
$$y: \quad \text{inc} \qquad \text{dec} \qquad \text{dec} \qquad \text{inc}$$

As we said, $x = -1$ is a singularity – the function is undefined there. Thus, the graph looks like this. (The vertical line at $x = -1$ is not part of the graph.)

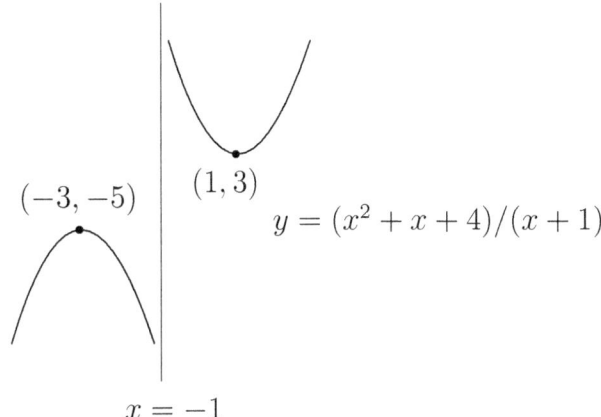

$(-3, -5)$ $(1, 3)$

$$y = (x^2 + x + 4)/(x + 1)$$

$x = -1$

∎

Problem T5.6. Draw a quick sketch of $y = x^4 - 8 \cdot x^3 + 18 \cdot x^2 - 20$.

Solution. The derivative has x as a factor:

$$y' = 4 \cdot x^3 - 24 \cdot x^2 + 36 \cdot x$$
$$= 4 \cdot x \cdot (x^2 - 6 \cdot x + 9)$$
$$= 4 \cdot x \cdot (x - 3)^2$$

The critical points $x = 0$ and $x = 3$. When $x = 3$, $y = 7$; when $x = 0$, $y = -20$. Here are the signs.

$$
\begin{array}{ccccc}
y' : & - & & + & + \\
& \rule{1cm}{0.4pt} & 0 & \rule{1cm}{0.4pt} & 3 \quad \rule{1cm}{0.4pt} \\
y : & \text{dec} & & \text{inc} & \text{inc}
\end{array}
$$

Can you draw the shape? There are horizontal tangents at $(0, -20)$ and $(3, 7)$.

∎

Using a Quick Sketch

We mentioned that there are several features of a graph that are apparent from its oscillations. One such feature is the number of *roots* that the function has – the number of times the graph of $y = f(x)$ crosses the x-axis.

Problem T5.7. How many real roots does $x^4 - 8 \cdot x + 3$ have?

Solution. Let $y = x^4 - 8 \cdot x + 3$, so that the polynomial has a root when the curve crosses the x-axis. The graph should tell! The derivative and its factors:

$$y' = 4 \cdot x^3 - 8 = 4 \cdot (x^3 - 2)$$

This has only one critical point: $x^3 - 2 = 0$ gives $x = \sqrt[3]{2}$. You may or may not remember how to factor $x^3 - 2$; let's show that we don't need to factor it. We can determine the signs of y' by choosing points on either side of $\sqrt[3]{2} \approx 1.26$ and observe the sign of y'. Taking $x = 1$ to the left of $\sqrt[3]{2}$, we see that $y' = -4$; thus, $y' < 0$ to the left of the cube root. Taking $x = 2$ to the right of $\sqrt[3]{2}$, we compute $y' = 24$, so that $y' > 0$ to the right of the critical point. This shows us that $y = x^4 - 8 \cdot x + 3$ has a minimum as $x = \sqrt[3]{2}$. Plugging in, we get $y \approx -4.56$. Since the point $(\sqrt[3]{2}, -4.56)$ is below the x-axis, and since y rises to the left and to the right[2] of $x = \sqrt[3]{2}$, we see that $x^4 - 8 \cdot x + 3$ has two real roots. ■

Here is another kind of problem for which a quick sketch is relevant.

Problem T5.8. Find the maximum and minimum of $y = -2x^3 + 3x^2 + 12x + 10$ for $0 \leq x \leq 3$.

Solution. Factor y':

$$-6x^2 + 6x + 12 = -6(x^2 - x - 2) = -6(x - 2)(x + 1)$$

---

[2]We assume you know that the values of a polynomial get large without bound as the x-values get large and positive or large and negative. For a polynomial of degree 4, y will get large and positive as x gets large in both directions.

and the sign diagram.

$$y' : \quad - \quad\quad\quad + \quad\quad\quad -$$
$$\underline{\quad\quad} \;\; -1 \;\; \underline{\quad\quad} \;\; 2 \;\; \underline{\quad\quad}$$

Now we consider the values of x between 0 and 3.

$$y' : \quad\quad + \quad\quad\quad -$$
$$0 \;\; \underline{\quad\quad} \;\; 2 \;\; \underline{\quad\quad} \;\; 3$$

The graph shows that y has a maximum at $x = 2$, where $y = 30$. The minimum of y on this interval must occur at one of the two endpoints. When $x = 0$, $y = 10$ and when $x = 3$, $y = 19$. The minimum occurs at $x = 0$. ■

Problem T5.9. Using the same function as the previous problem, find the maximum and minimum for $-2 < x \le 0$.

Solution. We use the same sign diagram for the critical points, superimposing the endpoints $-2, 0$.

$$y' : \quad\quad\quad - \quad\quad\quad\quad +$$
$$-2 \;\; \underline{\quad\quad} \;\; -1 \;\; \underline{\quad\quad} \;\; 0$$

We see that the minimum occurs at $x = -1$, $y = 3$. The maximum would have to be an endpoint. When $x = -2$, $y = 14$, and when $x = 0$, $y = 10$. Thus, $x = -2$ is the winner, but notice that $x = -2$ is not allowed in the specified interval $-2 < x \le 0$. It follows that *there is no maximum.* ■

Later in the course we will take an extended look at applied problems involving maximums and minimums.

The Second Derivative. The derivative measures whether a function is increasing or decreasing. The derivative of the derivative, called the *second derivative*, measures whether the derivative is increasing or decreasing. The second derivative of $f(x)$ is

$$\left[f'(x)\right]' \quad \text{denoted} \quad f''(x)$$

In terms of the graph, $f''(x)$ tells us the *concavity* of the curve. When $f''(x) > 0$, we say that $y = f(x)$ is *concave up*; when $f''(x) < 0$, we say that $y = f(x)$ is *concave down.* Between the two possibilities for the sign of $f'(x)$ and the two possibilities for the sign of $f''(x)$, there are four possibilities. The following circle picture gives the shape of the curve in each case.[3] We will discuss this in class.

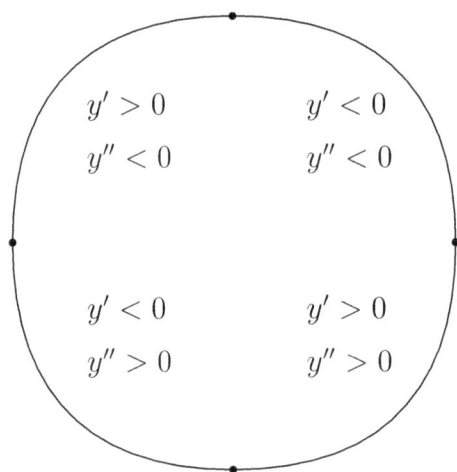

Problem T5.10. Graph $y = x \cdot \exp(-x)$ plotting critical points and showing oscillation and concavity.

Solution. We use the Product Rule and the Chain Rule to compute

$$y' = -(x - 1) \cdot \exp(-x)$$
$$y'' = (x - 2) \cdot \exp(-x)$$

The exponential function is always positive, no matter that its argument is $-x$, and so it doesn't contribute any signs. Here is the sign diagram, including

---

[3]We are using the arcs on the circle to indicate the shape. Of course, we are not saying that all curves are circular.

both derivatives. We abbreviate *concave down* to *dn* and *concave up* to *up*.

$$
\begin{array}{c|c|c|c}
y' : & + & - & - \\
y'' : & - & - & + \\
\hline
 & \underline{}\ 1\ \underline{}\ 2\ \underline{} \\
y : & \text{inc} & \text{dec} & \text{dec} \\
 & \text{dn} & \text{dn} & \text{up}
\end{array}
$$

And here is the graph; notice the change in shape at $x = 2$. (In drawing the picture, we have exaggerated that change somewhat to make it prominent.)

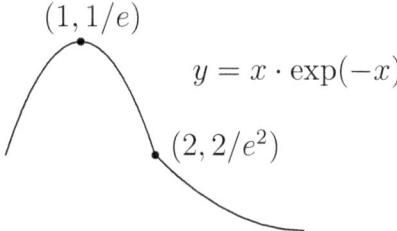

$(1, 1/e)$

$y = x \cdot \exp(-x)$

$(2, 2/e^2)$

Many functions that occur in economics have a positive derivative and a negative second derivative. Here is a made-up example: suppose your expected grade $G(h)$ on an exam depends on the number of hours h you study, according to the following table. (Grades are given on the 4-point scale, so 1=D, 2=C, etc.)

$$
\begin{array}{c|cccc}
h : & 0 & 1 & 2 & 3 \\
G(h) : & 1 & 2.75 & 3.5 & 3.75
\end{array}
$$

Your grade increases when h increases; that's hardly surprising: it says that the derivative $G'(h)$ is positive. But notice that when h changes from 0 to 1, $G(h)$ increases by 1.75, whereas when h changes from 1 to 2, $G(h)$ increases by only 0.75. The increase from the first additional hour of study is greater than the increase from the second hour. And for the third hour, the increase is only 0.25. It looks as if the derivative $G'(h)$ might be decreasing: $G''(h)$ is negative.

A function with a positive derivative and a negative second derivative exhibits *diminishing returns*; the *return* on investing an additional hour of study, in our example, is greater for the first hour than for the second hour, and greater for the second hour than for the third hour.

Problem T5.11. Let A, b be positive constants with $b < 1$. Show that the function $y = A \cdot x^b$ exhibits diminishing returns for $x > 0$.

Solution.   We use the Power Rule on x^b.

$$y' = A \cdot b \cdot x^{b-1} \quad \text{and} \quad y'' = A \cdot b \cdot (b-1) \cdot x^{b-2}$$

Since $x > 0$, the terms x^{b-1} and x^{b-2} are positive. Thus, $y' > 0$. Since $b < 1$, we have $b - 1 < 0$, and so $y'' < 0$. ■

2. Newton's Method

There are many ways to find numerical solutions to equations of the form $f(x) = 0$. We will discuss a recursive method called *Newton's Method* or the *Newton-Raphson Method*.[4]

We want to approximate a solution to the equation $f(x) = 0$. We start with a number R_0, and compute

$$R_{n+1} = R_n - \frac{f(R_n)}{f'(R_n)}$$

Of course, we need $f'(R_n) \neq 0$. This method is not infallible, but, remarkably often, if the initial R_0 is at all close to a solution, the sequence gives very accurate approximations to the solution.

Problem T5.12. Use Newton's Method to approximate $\sqrt{2}$.

Solution.   We need $\sqrt{2}$ to be a solution to an equation of the form $f(x) = 0$; let's use $x^2 - 2 = 0$, so that $f(x) = x^2 - 2$ in the notation of the method. Then

---

[4]As the names imply, the method goes back to Newton and/or Raphson, although the connection with calculus was, apparently, not elucidated at first.

$f'(x) = 2x$. Thus, the recursion is this:

(5.1)
$$R_{n+1} = R_n - \frac{R_n^2 - 2}{2R_n}$$

There is no single way to come up with the initial approximation R_0. Since $\sqrt{2}$ is between 1 and 2, we let $R_0 = 1$. Here is the sequence up to R_5.

$n:$	0	1	2	3	4	5
$R_n:$	1	1.5	1.4167	1.4142157	1.4142136	1.4142136

The R_n look constant from R_5 onward. Comparing to Excel's `sqrt(2)` (another approximation!), we see that we have apparent accuracy. ■

The recursion expression just used can be simplified significantly:

$$R_{n+1} = R_n - \frac{R_n^2 - 2}{2R_n} = R_n - \frac{R_n}{2} + \frac{1}{R_n} = \frac{R_n}{2} + \frac{1}{R_n}$$

You might recognize the resulting recursion as being an example of the Babylonian sequence! The Babylonians did not have Newton's Method in general, but their calculation of square roots was a special case of Newton's Method.

The expression we just obtained for the square root approximation is simpler than the recursion (5.1), and so it might be preferable. Such algebra is *sometimes* useful for employing Newton's Method, but we will not insist on it. When you use the method, feel free to use the expression $R_n - f(R_n)/f'(R_n)$ without simplifying.

Problem T5.13. The equation $e^x = c$ has solution $x = \ln(c)$. Use Newton's Method to approximate $\ln(c)$, where $c = 2, 3, 4, 5$.

Solution. Newton's Method needs an equation of the form $f(x) = 0$, and so we write the equation $e^x = c$ as $e^x - c = 0$. Thus, $f(x) = e^x - c$. Here is the recursion:

$$R_{n+1} = R_n - \frac{\exp(R_n) - c}{\exp(R_n)}$$

Taking $R_0 = 1$ in each case, here are some of the approximations.

$n:$	1	2	3	4	5
$\ln(2) \approx$	0.735759	0.694042	0.693148	0.693147	0.693147
$\ln(3) \approx$	1.103638	1.098625	1.098612	1.098612	1.098612
$\ln(4) \approx$	1.471518	1.389825	1.386301	1.386294	1.386294
$\ln(5) \approx$	1.839397	1.633963	1.609736	1.609438	1.609438

In each case, R_5 looks like a good approximation to the logarithm, as can be seen by checking the values in Excel. ■

We have introduced the recursion for Newton's Method; let's explain how it comes about. Imagine a graph $y = f(x)$ at a point $x = R_n$ near a root $f(c) = 0$. On the curve, this point is $(R_n, f(R_n))$. To find the root $x = c$, we would ideally like to follow the graph $y = f(x)$ to the point $(c, 0)$ on the x-axis. The idea behind Newton's Method is to use the tangent line to $y = f(x)$ at $x = R_n$ as a substitute for the curve. In other words, we find the point $(R_{n+1}, 0)$ where the tangent line meets the x-axis. (See the picture below.) We would want the two points $(R_n, f(R_n))$ and $(R_{n+1}, 0)$ both to be on the tangent line – that means that the slope of the tangent line can be found from these points:

$$\text{slope} = \frac{f(R_n) - 0}{R_n - R_{n+1}}$$

On the other hand, the slope is $f'(R_n)$, and so we have

$$f'(R_n) = \frac{f(R_n)}{R_n - R_{n+1}}$$

When we solve that equation for R_{n+1} we get the recursive equation that defines Newton's Method.

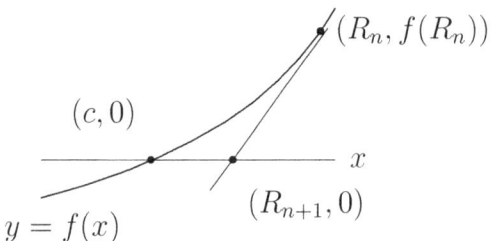

$(R_n, f(R_n))$

$(c, 0)$

x

$(R_{n+1}, 0)$

$y = f(x)$

Here is another example.

Problem T5.14. Find where the tangent line to $y = x^4 + x^2 + x + 3$ is horizontal.

Solution. For the tangent line to be horizontal, its slope needs to be 0. Thus, we want to solve $dy/dx = 0$:

$$4 \cdot x^3 + 2 \cdot x + 1 = 0$$

We don't know how to factor this cubic; graphing it shows that it has one root; we use Newton's Method:

$$R_{n+1} = R_n - \frac{4 \cdot R_n^3 + 2 \cdot R_n + 1}{12 \cdot R_n^2 + 2}$$

Starting with $R_0 = 0$, we compute $R_6 \approx R_7 \approx -0.3855$. This gives us the x-value, and the y-value ≈ 2.7852. ∎

3. The Chain Rule Revisited

3.1. Differentials. It is interesting to write the Chain Rule in terms of differentials. Let $y = f(x)$ and $x = g(t)$, so that $y = f(g(t))$. Here are the various derivatives:

$$\frac{dy}{dx} = f'(x) \qquad \frac{dx}{dt} = g'(t) \qquad \frac{dy}{dt} = \big[f(g(t))\big]'$$

Here is the Chain Rule, with $x = g(t)$ substituted in.

$$\big[f(g(t))\big]' = f'(g(t)) \cdot g'(t) = f'(x) \cdot g'(t)$$

We replace the derivatives on the far left and far right by differential expressions, and we get the differential version of the Chain Rule.

Chain Rule – Differential Version Given $x = g(t)$ and $y = f(x)$, where $g(t)$ and $f(x)$ are differentiable, we have

(5.2)
$$\frac{dy}{dt} = \frac{dy}{dx} \cdot \frac{dx}{dt}$$

∎

Equation (5.2) *looks like* the dx factors were canceled. When we introduced differentials, we said we would not give an explicit definition – in any case they are not numbers, and so the dx factors don't cancel *literally*. But the Chain Rule does say that the equation (5.2) is true, so the point is that we can apparently manipulate differentials as if they were numbers. We claimed before that differentials would prove useful; the Chain Rule justifies that promise.

3.2. Inverse Functions. We will consider a very simple example: $y = x^5$. The differential notation dy/dx asks for the derivative of y as a function of x. Thus,

$$\frac{dy}{dx} = 5 \cdot x^4$$

We can rewrite the equation $y = x^5$ by taking the fifth root: $y^{1/5} = x$. In this equation x is a function of y, and the derivative dx/dy would ask for the derivative of that function.

$$\frac{dx}{dy} = \frac{1}{5} \cdot y^{-4/5}$$

This derivative is defined as long as $y \neq 0$, since the negative exponent indicates a term in the denominator.

The Chain Rule showed that differentials seem to behave algebraically. Let's show that they behave that way in this context. Compute

$$
\begin{aligned}
\frac{dy}{dx} \cdot \frac{dx}{dy} &= 5 \cdot x^4 \cdot \frac{1}{5} \cdot y^{-4/5} \\
&= x^4 \cdot y^{-4/5} && \text{canceling 5's} \\
&= x^4 \cdot (x^5)^{-4/5} && \text{using that } y = x^5 \\
&= x^4 \cdot x^{-4} && \text{multiplying consecutive exponents} \\
&= x^{4-4} = x^0 = 1 && \text{adding exponents}
\end{aligned}
$$

Here is the upshot:

(5.3) $$\frac{dy}{dx} \cdot \frac{dx}{dy} = 1$$

As with the equation (5.2), it looks as if the differentials are playing algebra – we keep saying that is why they are useful.

The function x^5 and $y^{-1/5}$ are called *inverse functions*; they undo each other:

$$(x^5)^{1/5} = x^{5/5} = x^1 = x \quad \text{and} \quad (y^{1/5})^5 = y^{5/5} = y^1 = y$$

You have seen this pattern before: $y = \exp(x)$ if and only if $\ln(y) = x$. And it occurs in other situations. Let's show that (5.3) holds whenever we have differentiable inverse functions. If $y = f(x)$ and $x = g(y)$ are inverses, then

$$x = g(y) = g(f(x))$$

Using the ordinary Chain Rule on both sides:

$$1 = g'(f(x)) \cdot f'(x)$$

Now differentials: $g'(f(x)) = g'(y) = dx/dy$, and $f'(x) = dy/dx$. We see that (5.3) holds.

But there is an additional fact. It turns out that equation (5.3) holds even if we do not have an explicit formula for the inverse function, as long as the derivatives involved in (5.3) are not 0. This fact is called the *Inverse Function Theorem*.[5] Here is an applied example.

Problem T5.15. We observe that demand x for pianos depends on price p (thousands of dollars). Suppose that

$$x = \frac{44 \cdot 10^3}{p^3} + \frac{30}{p} + 3$$

An economist might like to think of price as a function of demand. What is dp/dx when $p = 10$? (We will see later that this derivative is called the *marginal price*.)

---

[5]Technically, if $y = f(x)$ has a non-zero derivative on an open interval, then the y values will traverse an open interval, as well, and there will be an inverse function $x = g(y)$ for y in its open interval.

Solution. The definition of the expression dp/dx requires us to find $p = f(x)$, and then $dp/dx = f'(x)$. We have an equation x as a function of p, but it is not clear that we can solve for p in this equation. The Inverse Function Theorem tells us that we don't have to! Using the equation analogous to (5.3) in the present context, we see that

$$\frac{dp}{dx} \cdot \frac{dx}{dp} = 1$$

and this equation can be written

(5.4) $$\frac{dp}{dx} = \frac{1}{dx/dp}$$

as long as $dx/dp \neq 0$. We use the equation for x as a function of p to compute dx/dp, remembering that $1/p^3 = p^{-3}$.

$$\frac{dx}{dp} = \left[\frac{44 \cdot 10^3}{p^3} + \frac{30}{p} + 3\right]' = -\frac{3 \cdot 44 \cdot 10^3}{p^4} - \frac{30}{p^2}$$

When $p = 10$, we see that

$$\frac{dx}{dp} = -\frac{3 \cdot 44 \cdot 10^3}{10^4} - \frac{30}{10^2} = -13.5 \text{ pianos per thousand dollars}$$

Equation (5.4) shows that

$$\frac{dp}{dx} = \frac{1}{dx/dp} = -\frac{1}{13.5} \text{ thousand dollars per piano}$$

■

Notice that both derivatives were negative – increasing price decreases demand, and so increasing demand decreases price.

We can use this reasoning to derive the formula for $(\ln(x))'$ from the formula for $(e^x)'$. Let $y = \ln(x)$, so that $x = e^y$. Then

$$\left[\ln(x)\right]' = \frac{dy}{dx} = \frac{1}{dx/dy} = \frac{1}{(e^y)'} = \frac{1}{e^y} = \frac{1}{x}$$

Previously, we stated that $\ln'(x) = 1/x$, but we didn't explain where that formula came from. Now we see that it comes from the formula $\exp'(x) = \exp(x)$ and the Inverse Function Theorem.

Here is a problem that combines (5.2) and (5.3).

Problem T5.16. A particle traces a curve in the xy-plane. At time t, we have $x = (t^2 - 1)/(t^2 + 1)$ and $y = 2t/(t^2 + 1)$. What is the slope of the path of the particle when $t = 2$? (Slope is dy/dx, as always!)

Solution. We see that we can calculate dx/dt and dy/dt. Treating differentials algebraically, and not worrrying about reasoning for the moment, we might expect that

$$\frac{dy}{dx} = \frac{dy/dt}{dx/dt}$$

provided that $dx/dt \neq 0$. Compute dy/dt using the Quotient Rule.

$$\frac{dy}{dt} = \frac{2(t^2 + 1) - 2t(2t)}{(t^2 + 1)^2} = \frac{2 - 2t^2}{(t^2 + 1)^2}$$

Similarly,

$$\frac{dx}{dt} = \frac{2t(t^2 + 1) - (t^2 - 1)2t}{(t^2 + 1)^2} = \frac{4t}{(t^2 + 1)^2}$$

Then

$$\frac{dy}{dx} = \frac{dy/dt}{dx/dt} = \frac{(2 - 2t^2)/(t^2 + 1)^2}{4t/(t^2 + 1)^2} = \frac{2 - 2t^2}{4t}$$

When $t = 2$, we would have $dy/dx = -6/8 = -3/4$. That would be the slope of the path when $t = 2$. ∎

We worked the previous problem in a typical way – people don't usually worry too much about justifying differential calculations. For completeness, let's show that what we did is valid. We had $dx/dt = 4t/(t^2 + 1)^2$, and that's not zero when $t = 2$. The Inverse function Theorem then says that t is a function of x near that point, and (5.3) kicks in to show that

$$\frac{dt}{dx} = \frac{1}{dx/dt} = \frac{(t^2 + 1)^2}{4 \cdot t}$$

Theoretically, t is a function of x, and the equation $y = 2t/(t^3+1)$ shows that y is a function of t. Then y is a function of x and the Chain Rule shows that

$$\frac{dy}{dx} = \frac{dy}{dt} \cdot \frac{dt}{dx}$$

This shows that our calculation of dy/dx was justfiied.

As we move forward, we won't stop to anaylze the reasoning behind our use of differentials. As long as we don't divide by zero derivatives, the algebraic-like equations (5.2) and (5.3) will win through.

4. Marginals

Suppose that the cost C of making something is a function of the number x made. Not unreasonable. If C, x have specific values, the *marginal cost* is the change in cost when x increases by 1. In other words, marginal cost is ΔC when $\Delta x = 1$. Since

$$\frac{\Delta C}{\Delta x} \approx \frac{dC}{dx}$$

we have $\Delta C \approx dC/dx$ when $\Delta x = 1$. Because the derivative dC/dx can often be computed, that derivative is usually taken for the marginal cost.[6] We might only be interested in the sign of dC/dx, since that tells us whether C will increase or decrease if x increases.

If M is a quantity (particularly, an economic quantity) that depends on the level x of production of something, then *marginal M* is the change in M when the one extra unit of production – generally taken to be dM/dx. There is sometimes a little insignificant confusion about the units of marginal M. For instance, if M is dollars and x is pianos, then dM/dx is *dollars per piano* – a unit cost. However, we are taking dM/dx as an approximation for ΔM, and ΔM would be just dollars. This distinction is usually immaterial; we will use the units of the derivative to be consistent with our previous terminology.

---

[6]Alternatively, some sources *define* marginal cost to be dC/dx, but in any case the idea is to use the derivative to estimate ΔC.

If we remember that ΔM was actually $\Delta M/\Delta x$, with $\Delta x = 1$, then we arrive at the derivative units: dollars per piano.

Problem T5.17. The cost of producing x units of a good is the sum of three quantities: a constant overhead of \$6000 for machinery, a production cost of \$50 per unit, and a quantity discount \$1000 times the square root of x. If we currently produce 120 units, what is the marginal cost?

Solution. The description of the cost shows that

$$C = 6000 + 50 \cdot x - 1000 \cdot \sqrt{x}$$

The marginal cost is none other than dC/dx:

$$\frac{dC}{dx} = 50 - 1000 \cdot \frac{1}{2} \cdot x^{-1/2} = 50 - \frac{500}{\sqrt{x}}$$

When $x = 120$, we have $C' \approx 4.36$. We expect the cost to increase by roughly \$4.36 if one extra unit is produced. ■

Problem T5.18. In the previous problem, the total value is $x \cdot C$. What is the marginal total value when 100 units are produced?

Solution. The total value $T = x \cdot C$ where C is cost. By the Product Rule,

$$\frac{dT}{dx} = C + x \cdot C'$$

We have the formula for C' in the previous problem. Plugging in $x = 100$, we compute that $T' \approx 1000$ (dollars per unit). ■

The word *marginal* can indicate various other sorts of derivatives. Suppose, for example, that the production P is determined by the amount of labor L used; let's write $P = f(L)$. The derivative

$$\frac{dP}{dL} = f'(L)$$

is called the *marginal product of labor*. The abstract phrase *marginal Y of X* is simply dY/dX.

Here is a typically convoluted problem. Algebra and calculus sort everything out.

Problem T5.19. It takes 3 hours of labor to produce one table. The production cost of x tables is $100x - x^2/10$ dollars. What is the marginal production cost of hours of labor when we make 10 tables?

Solution. Let L be the number of hours of labor, and let V be the production cost. The marginal production cost of labor is simply dV/dL, and that's what we want.

The problem statement tells us that $V = 100x - x^2/10$, and since it takes 3 hours to make one table we have $L = 3x$. Thus, we can easily get these derivatives:

$$\frac{dV}{dx} = 100 - \frac{2}{10}x = 100 - x/5 \quad \text{and} \quad \frac{dL}{dx} = 3$$

The derivative we want can be found using the differentials:

$$\frac{dV}{dL} = \frac{dV/dx}{dL/dx} = \frac{100 - x/5}{3}$$

Now we are ready for specific numbers. With $x = 10$, the marginal value of labor is

$$\frac{100 - 10/5}{3} = \frac{98}{3} \text{ dollars per hour}$$

∎

Let's interpret the answer, 98/3 dollars per hour, of the previous problem. Given that we now make 10 tables (using 30 hours of labor), if could increase labor by one hour, our production costs would go up by roughly 98/3 dollars.

Problem T5.20. Suppose that the supply S of pencils (sold in boxes of 1000) satisfies $S = 2 + 1.3 \cdot p$, where p is the price of a box in dollars. Suppose that the demand D for pencils satisfies $D = 35 - m \cdot p$, where m is a constant. Describe m as a marginal. If m increases, what happens to the price where supply and demand are equal? What happens to revenue when $m = 2$ and increases?

Solution. We see that $dD/dp = -m$, so that $-m$ is the marginal demand of price.

We are interested in the price where $S = D$; let's call that price q. (We use q rather than p, since q is a specific value of the variable p.) Using the expressions for S and D, we have

$$2 + 1.3 \cdot q = 35 - m \cdot q$$

Moving the q-terms together, this is

$$1.3 \cdot q + m \cdot q = 35 - 2 \quad \text{which is} \quad (1.3 + m) \cdot q = 33$$

And we have

$$q = \frac{33}{1.3 + m}$$

To measure the change in q as m increases, we take the derivative

$$\frac{dq}{dm} = \left(33 \cdot (1.3 + m)^{-1}\right)' = -33 \cdot (1.3 + m)^{-2} = -\frac{33}{(1.3 + m)^2}$$

Since the derivative is negative, we see that q *decreases* as m increases.

As for revenue, that is $R = D \cdot p$. When $p = q$, we get this formula for R:

$$R = D \cdot q = (35 - m \cdot q) \cdot q = 35 \cdot q - m \cdot q^2$$

We can take the derivative with respect to m, remembering that we already have a formula for dq/dm.

$$\frac{dR}{dm} = 35 \cdot \frac{dq}{dm} - 1 \cdot q^2 - m \cdot 2q \cdot \frac{dq}{dm}$$

When $m = 2$, we see that $q = 33/3.3 = 10$ and $dq/dm = -33/3.3^2 \approx -3.03$, so that

$$(5.5) \qquad \frac{dR}{dm} \approx 35 \cdot (-3.03) - 100 - 40 \cdot (-3.03) = -84.85$$

If m increases, R will decrease dramatically. ∎

In the previous problem, suppose it would cost us \$7 to decrease m by 0.1. (Perhaps, advertising would make demand less sensitive to price.) Should we spend the $7? The formula (5.5) estimates

$$\Delta R \approx -84.85 \cdot \Delta m = -84.85 \cdot (-0, 1) = 8.485$$

We can increase revenue by $8.49 by spending $7; seems like a good idea.

There are many terms involving marginals. It is not the purpose of this course to learn the terminology, but here is a typical problem.

Problem T5.21. Suppose that $p = 1000 - x^{1/3}$, with price p and demand x for a certain good. The *price elasticity of demand* is the ratio of relative change of demand to relative change of price. Find that elasticity when $x = 100$.

Solution. The set-up has x as independent variable, and so the relative change in p is

$$\frac{1}{p} \cdot \frac{dp}{dx} = \frac{1}{p} \cdot (-1) \cdot \frac{1}{3} \cdot x^{-2/3} = -\frac{1}{3 \cdot p \cdot x^{2/3}}$$

The relative change in x is

$$\frac{1}{x} \cdot \frac{dx}{dx} = \frac{1}{x}$$

The elasticity is the ratio of the relative changes (prepare for some algebra):

$$\frac{1/x}{-1/(3 \cdot p \cdot x^{2/3})} = -\frac{3 \cdot p}{x^{1/3}}$$

When $x = 100$, we get $p \approx 995.36$, and the elasticity is roughly -625. ■

CHAPTER 6

Linear Optimization

An *optimization problem* seeks the maximum or minimum value of a function. For a simple function of one variable, this information can be obtained from the graph – from the signs of the derivative. More practical optimization problems usually involve many variables, along with complicated conditions on those variables that make the one-variable graphing approach impractical. Among the more complicated problems there is a fundamental dichotomy between *linear optimization problems*, which are introduced in this chapter, and the *non-linear optimization problems* introduced in Chapter 12 (in the second term of this course). We will use simple examples to introduce linear problems; once we understand the basic form and terminology, we will consider more complicated (interesting!) problems. We will also study decision-making based on the properties of solutions.

We will be able to use graphing to handle simple examples; later we will be using Excel to obtain numerical solutions.

1. Simple Examples

For our first example we will focus strictly on the mathematical details of a typical, small problem.

Problem T6.1. Find the maximum of $Z = 6 \cdot x + 7 \cdot y$, where $3 \cdot x + 5 \cdot y \leq 15$ and $2 \cdot x + y \leq 4$ and $x, y \geq 0$.

Solution. The quantity Z being optimized (in this case, maximized) is called the *objective*.[1] The objective in this problem is a function of the variables x, y; those are the *problem variables*. The other conditions on the variables are the *constraints*.

The constraints may look complicated, but they are easy to graph. To graph the inequality $3 \cdot x + 5 \cdot y \leq 15$, we solve for y:

$$3 \cdot x + 5 \cdot y \leq 15 \quad \text{yields} \quad y \leq \frac{1}{5} \cdot (15 - 3 \cdot x) = 3 - \frac{3}{5} \cdot x$$

The inequality $y \leq 3 - \frac{3}{5} \cdot x$ refers to points on or below the line $y = 3 - \frac{3}{5} \cdot x$. Similarly, the inequality $2 \cdot x + y \leq 4$ comes to $y \leq 4 - 2 \cdot x$, and that's the points on or below the line $y = 4 - 2 \cdot x$. The other two constraints $x, y \geq 0$ say we are in the first quadrant. Here, to the left, is a picture of the constraints; we included the x-intercepts of the two lines, as well as the y-intercepts.

The problem requires all the constraints to hold. That means that the constraints define the set of points on and inside the small quadrilateral defined by the picture; that quadrilateral is pictured to the right. We have caculated the intersection point between the two lines. We claim that the maximum of Z cannot occur interior to the quadrilaterial. This is easy to see: at an interior point, we can increase x or y, or both, and $Z = 6 \cdot x + 7 \cdot y$ will increase.

---

[1]Notice that this use of the word *objective* is not the usual one. Usually, an objective is a goal. The word *objective* in an optimization problem refers to the quantity being optimized, as opposed to the desire to optimize that quantity. This possibly confusing usage has become standard.

We also claim that the maximum of Z occurs at one of the corner points of the quadrilateral. This will best be shown in class: briefly, along each boundary line we can write Z as a linear function of one variable. A linear function has a maximum at one of its endpoints – the endpoints of the boundary lines are the corners in the picture. We will give more specific details in class.

To find the maximum of Z, we have only to calculate its values at each corner:

$$\begin{array}{ccccc} \text{point:} & (0,0) & (2,0) & (0,3) & (5/7, 18/7) \\ Z & 0 & 12 & 21 & 156/7 \end{array}$$

Because $156/7 \approx 22.3$, we see that the maximum of Z is $156/7$, occuring at $(5/7, 18/7)$. ∎

In the problem just worked, the point $(5/7, 18/7)$ gives values of the problem variables: $x = 5/7$ and $y = 18/7$. We call this a *solution* to the problem. To repeat: the solution gives the values of the problem variables, not the value of the objective. The value of the objective $Z = 156/7$ at the solution is the maximum of the objective.

Here is a similar problem that we will work in class.

Problem T6.2. Find the maximum of $Z = 3 \cdot a + 2 \cdot b$ where $2 \cdot a + 7 \cdot b \leq 14$ and $7 \cdot a + 5 \cdot b \leq 35$ and $a, b \geq 0$. ∎

In each of the two problems we considered, the objective (quantity) is a *linear function* of the problem variables. Also, the constraints involve linear expressions in the problem variables. In the examples we used, the constraints were non-strict inequalities. They can also be equations, as we will see in more complicated examples. The presence of only linear expressions in the objective and constraints is what makes an optimization problem *linear*.

It is possible for the constraints of a linear optimization problem to contradict each other. In that case, we say that the problem is *infeasible*. It may seem silly to consider constraints that don't make sense, but that can easily happen if we try to impose a complicated set of conditions on an applied problem. If there are values of the problem variables that satisfy the constraints, then the problem is *feasible*. A feasible problem may or may not have a solution.

Problem T6.3. Can we find the maximum of $Z = x$ where $x + y = 7$ and $x \geq 0$?

Solution. The problem is feasible, since $x = 0, y = 7$ satisfies the constraints. But the problem has no *solution*, since the objective Z has no maximum. We can make $Z = x$ as large as we wish and have $y = 7 - x$ to keep the constraints happy. (Note that we are allowed to let y be negative.) ∎

A feasible problem with a solution may have more than one solution.

Problem T6.4. Minimize $Z = -3 \cdot x - 4 \cdot y$ where $6 \cdot x + 8 \cdot y \leq 48$ and $x, y \geq 0$.

Solution. As before, we graph the constraints. Remembering that $x, y \geq 0$, the line $6 \cdot x + 8 \cdot y = 9$ runs between $(0, 9/8)$ and $(3/2, 0)$. The value of Z is $-9/2$ at both the corner points, and it is 0 at $(0, 0)$. Thus, the minimum of Z is $-9/2$, occurring at $(0, 9/8)$ and $(3/2, 0)$. Actually, $Z = -9/2$ at all points on the line segment joining $(0, 9/8)$ and $(3/2, 0)$. ∎

2. More Complicated Examples

Typical optimization problems involve many variables and constraints, and so we cannot hope to solve general problems by graphing. We hope that graphing simple problems gives you some idea what to expect in general. Moving to more realistic problems, we will emphasize *setting up* problems, and then we will use Excel to find numerical answers.

We recommend the book [**12**], because it is an excellent source for a wide array of application problems and because it teaches an overall approach to linear problems. The author, George Dantzig, discovered a fundamental numerical algorithm for solving linear optimization problems, and his algorithm is the engine of many software solvers – for instance, the Excel `Solver`. We will use the `Solver` to obtain numerical answers; in class we will demonstrate this.

2.1. Mixing. There are a host of linear optimization problems that involve mixing ingredients together to make something: choosing foods to construct a diet, choosing metals to manufacture an alloy, etc. In the problems we consider there are many ways to mix the ingredients – that's probably a surprising feature, since you are used to situations where there is one recipe for the desired mix. (e.g. There is only one way to make a water molecule out of hydrogen and oxygen.)

Problem T6.5. We manufacture four models of desks.[2] Each desk is constructed in the carpentry shop and then sent to the finishing shop. The number of hours of labor required for each model at each stage is given in the table below. We also list the per desk profit realized from the sale of each model. How many desks of each type should we make to maximize profit, if we have up to 6000 hours available for carpentry and 4000 for finishing? (Note: the *mix* here is the mix of numbers of models scheduled to be constructed.)

	model 1	model 2	model 3	model 4
carpentry	4	9	7	10
finishing	1.5	1.5	3	40
unit profit	12	20	18	40

Solution. The objective is profit Q, and it is a function of the number of each desk model we make. Thus, we have four variables: M_1, M_2, M_3, M_4 giving us the number of each model made. Then profit is easily computed:

$$Q = 12 \cdot M_1 + 20 \cdot M_2 + 18 \cdot M_3 + 40 \cdot M_4$$

---

[2]Modified from an example in [**12**, p.50]

Notice that we know the objective profit Q as a *function* – we don't have specific numerical values of Q.

Constraints? The variables are non-negative. We have up to 6000 hours of carpentry labor; reading the table we see the number of hours for each model:

$$4 \cdot M_1 + 9 \cdot M_2 + 7 \cdot M_3 + 10 \cdot M_4 \leq 6000$$

We pause to mention that inequality constraints are very common in optimization problems. We might assume that if we want to maximize profit, we will need to use all 6000 hours here, and so we would actually have an equation rather than an inequality. That may be the case; the solution will disclose the situation. When a problem has many constraints, they may not all be able to produce equality in every inequality.

The finishing hours constraint is similar to the constraint for carpentry.

$$1.5 \cdot M_1 + 1.5 \cdot M_2 + 3 \cdot M_3 + 40 \cdot M_4 \leq 4000$$

The Excel `Solver` gives this solution to maximize Q:

$$M_1 \approx 1379, \quad M_2 \approx 0, \quad M_3 \approx 0, \quad M_4 \approx 42.28, \quad Q \approx 18482.76$$

We might wonder about the decimal answer for M_4 in the last problem. Can we make fractional numbers of desks? In general, it is a significant complication to require integer values in an optimization problem. The `Solver` allows us to constrain some or all of the variables to be integers, and, if the problem is not too large, then the `Solver` will find the constrained solution. However, for larger problems the `Solver` can work for a long time without getting a solution, and there's no way to tell in advance how long it will take. When we constrain the desk problem variables to be integers, we get $M_1 = 1380$ and $M_4 = 48$ and $M_2 = M_3 = 0$, and the maximum of Q is 18480. Note that this solution *is not obtained* merely by rounding the decimal values in the first solution – that just does not work in general. In this text, we do not want to introduce the many nuances possible when variables have to be integers, and so we will allow decimal answers without worrying. But you should be aware that there is a major issue here. We also remind you that we are dealing with *models*, and the answer, decimal or integer, might be an approximation in any case.

2.2. Transportation. We imagine a single good – say pianos. We have a bunch of warehouses, with varying numbers of pianos in each. And we have store locations from which various numbers of pianos will be sold. We want to ship pianos from the warehouses to the stores at minimal cost.

Let's suppose we have 3 warehouses and 4 stores, with unit shipping costs from each warehouse to each store indicated in the following table.[3] The last column of the table shows how many pianos must be shipped to each store. The bottom row of the table gives the supply of pianos at each warehouse.

warehouse:	A	B	C	demand
store 1:	10	50	20	60
store 2:	5	10	30	70
store 3:	20	10	6	120
store 4:	30	30	30	100
supply:	100	200	80	–

The key to this problem is to realize that we need 12 variables: one variable for each warehouse, store pair; the value of the variable will give the number of pianos shipped from the particular warehouse to the particular store. Let $X[A1]$ be the number shipped from warehouse A to store 1; let $X[A2]$ be from warehouse A to store 2, and so on. The objective is the cost Z:

$$Z = 10 \cdot X[A1] + 5 \cdot X[A2] + \cdots + 6 \cdot X[C3] + 30 \cdot X[C4]$$

The coefficients come from the cost table. We have omitted some of the terms; there are 12 of them altogether!

---

[3]Shipping costs can vary due to many factors such as distance.

As for constraints, each store needs to receive the correct number of pianos. Store 1 needs 60 pianos, and so

$$X[A1] + X[B1] + X[C1] = 60$$

There is a similar constraint for the other three stores. Also, each warehouse can send no more pianos than it has in stock. Warehouse A has 100 pianos, and so

$$X[A1] + X[A2] + X[A3] + X[A4] \leq 100$$

Notice the inquality here. There are 380 pianos, total, at the warehouses, and the stores want a total of 350, so not all pianos will be shipped. We get three warehouse constraints. Finally, all the variables are non-negative, since we cannot ship negative numbers of pianos!

In the `Solver`, the variables will be laid out in a 4×3 table, corresponding to the shipping cost table, as we will see in class. The minimum cost is $Z = 4980$.

2.3. Time Sequence. We consider a problem in which there is a sequence of variable values in time. For instance, each week we produce a certain amount of a quantity. The problem we consider is a step toward more realistic (larger!) problems.

Problem T6.6. During a given week, over a four week period, we have workers and trainees. It takes one week to turn a trainee into a worker. Workers are divided into producers (making a product) and those who are idle. Here is a table of revenue produced for each producer, the cost per idle worker, and the cost per trainee. There is a maximum number of producers for each week in the table, as well. Assuming we begin with 10 workers, figure out how many producers, idle, and trainees we need each week to maximize total revenue.

week:	1	2	3	4
producer (unit revenue)	12	10	10	12
idle (unit cost)	5	4	2	4
trainee (unit cost)	6	7	4	-
maximum producers	8	30	25	40

Solution. As we have been advocating, we turn everything into a variable! We'll have workers W_n during weeks $1, 2, 3, 4$, and trainees T_n during weeks $1, 2, 3$. (Because there are only four weeks, it doesn't make sense to pay for trainees during week 4.) We'll have producers P_n for weeks $1, 2, 3, 4$ and idle I_n for weeks $1, 2, 3, 4$. The objective R is revenue: production minus costs.

$$R = 12 \cdot P_1 + 10 \cdot P_2 + 10 \cdot P_3 + 12 \cdot P_4$$
$$- 5 \cdot I_1 - 4 \cdot I_2 - 2 \cdot I_3 - 4 \cdot I_4$$
$$- 6 \cdot T_1 - 7 \cdot T_2 - 4 \cdot T_3$$

Even though this is complicated, it is just a `sumproduct` expression in Excel. We want to maximize R as a function of all these variables.

For constraints, we just read through the problem carefully. Trainees turn into workers the next week. Thus,

$$W_2 = W_1 + T_1 \qquad W_3 = W_2 + T_2 \qquad W_4 = W_3 + T_3$$

These equations are called *time sequence equations*, since they show how the variables step through the time intervals. Next, workers are divided into producers and idle.

$$W_1 = P_1 + I_1 \qquad W_2 = P_2 + I_2$$
$$W_3 = P_3 + I_3 \qquad W_4 = P_4 + I_4$$

These equations are sometimes called *material balance equations*, since they show how various groups are divided up – they provide a kind of accounting of the workers. It is not crucial to use the terms *times sequence equations* and *material balance equations*, but we think they help in setting up the more complicated problems. Time sequence equations show how to get from one time step to the next one; material balance equations show how items are categorized at a single time step.

There is a maximum number of producers each week:

$$P_1 \le 8 \qquad P_2 \le 30 \qquad P_3 \le 25 \qquad P_4 \le 40$$

And we begin with $W_1 = 10$. As usual, all the variables are non-negative, and we have our problem set-up!

In class, we will describe using Excel to solve this problem. Here is a table showing numbers of workers each week.

week:	1	2	3	4
workers:	10	30	30	40

Our solution shows that we will be training workers during weeks 1 and 3, and we will have idle workers during those weeks, as well. ■

3. Shadow Prices – Lagrange Multipliers

Linear optimization problems contain a multitude of coefficients; some of these values may be negotiable. Let's go back to the first problem worked in this chapter and express one of the coefficients as a parameter that can be changed.

Problem T6.1 – revisited.   Find the maximum of $Z = 6 \cdot x + 7 \cdot y$, where $3 \cdot x + 5 \cdot y \le A$ and $2 \cdot x + y \le 4$ and $x, y \ge 0$. The parameter A starts at 15, but it might change; what happens to the maximum value of Z in that case? ■

Solution.   We have replaced the right side 15 in the first constraint by a variable right side A. Currently $A = 15$, but we want to know what will happen if A changes. Notice that the objective Z is a function of the problem variables x, y; it doesn't mention A at all. The change in A doesn't affect the objective Z, it affects the *maximum of Z*. Let's see this.

The solution to this problem occurs where the two lines meet; we solve those equations using our parameter A.

$$3 \cdot x + 5 \cdot y = A \quad \text{and} \quad 2 \cdot x + y = 4 \quad \text{gives} \quad x = \frac{20 - A}{7}, \; y = \frac{2 \cdot A - 12}{7}$$

This solution gives maximum objective

$$\bar{Z} = \frac{8 \cdot A + 36}{7}$$

We have used the notation $\bar{Z}$ to distinguish the maximum value of the objective from Z as a function of x, y. Notice that $\bar{Z}$ is a function of the parameter A. When $A = 15$, we get the solution originally obtained: $x = 5/7$ and $y = 18/7$ and $\bar{Z} = 156/7$.

It is easy to use our formula for $\bar{Z}$ to get information about $\Delta\bar{Z}$. Since $\bar{Z}$ is a linear function of the parameter A, we see that

$$\frac{\Delta\bar{Z}}{\Delta A} = \frac{d\bar{Z}}{dA} = \frac{8}{7}$$

If A increases from 15 to 16, so that $\Delta A = 1$, then

$$\frac{8}{7} = \frac{d\bar{Z}}{dA} = \frac{\Delta\bar{Z}}{\Delta A} = \Delta\bar{Z}$$

We expect the maximum objective to increase by $8/7$. ∎

Let's call attention to the shift in point of view here. While we are working the optimization problem the objective Z is a function of the problem variables x, y, and A is the constant 15. In other words, A is fixed while we do the optimization problem. Once we do the problem, the maximum objective $\bar{Z}$ is a function of the parameter A, without the problem variables.

The derivative $d\bar{Z}/dA = 8/7$ is a special case of the derivative of the optimal objective with respect to a parameter. This derivative is identified by several technical terms, for it arises in several contexts. Mathematicians and practitioners of some other disciplines call it a *Lagrange multiplier*. In economics problems, it may be called a *shadow price* – we will explain that term momentarily. People also say that the derivative measures the *sensitivity* of the maximum with respect to the parameter A. We will use the term *Lagrange multiplier*, while nodding to other terms as they make sense.

Here is a typical applied problem.

Problem T6.7. We make (cheap) chairs and tables. It costs \$10 to make each chair, and \$50 to make each table, and we can spend up to \$500 to make them. Each chair takes 3 hours of labor and each table 2 hours; we have 60 hours of labor available. We get \$22 profit from each chair and \$35 from each table. How many chairs and tables should we make to maximize profit? What would it be worth to us to increase the available hours of labor by 1?

Solution. Let x be the number of chairs we make, and let y be the number of tables. Cost: $10 \cdot x + 50 \cdot y \le 500$. Labor: $3 \cdot x + 2 \cdot y \le 60$. Also, $x, y \ge 0$. The objective is profit $P = 22 \cdot x + 35 \cdot y$.

Let's think about the second question: the question about what happens if the amount of labor changes. The amount of labor L is a parameter, currently 60. We see that the objective profit P is a function of the variables x, y; the profit does not mention the parameter L. The *maximum value* $\bar{P}$ of P, however, will be a function of L. The Lagrange multiplier would be $d\bar{P}/dL$.

The `Solver` finds the solution

$$x = 15.4, \quad y = 6.9, \quad \text{so that} \quad \bar{P} = 580.8$$

When the `Solver` reports that it has found a solution, the right side of the dialogue box shows three options: `Answer` and `Sensitivity` and `Limits`. If you select `Sensitivity`, then a new worksheet will be produced that shows the Lagrange Multipliers (or it may call them shadow prices) – one multiplier for each constraint. Write L for the amount of labor – currently 60 hours. The Lagrange Multiplier for L goes with the constraint $3 \cdot x + 2 \cdot y \le L$. Excel calculated the Lagrange multiplier for the labor constraint to be 5.8. Thus,

$$\frac{d\bar{P}}{dL} = 5.8$$

An additional hour of labor is $\Delta L = 1$. Then $\Delta \bar{P} \approx 5.8$; an additional hour of labor brings in an additional profit around \$5.80. That's how much the additional hour of labor is worth to us. ∎

Here is a decision-making problem arising from what we just did: suppose we could increase the available labor hours by one hour for only \$5.00. The shadow price (Lagrange Multiplier) estimates an increase of \$5.80 in profit, so we come out ahead by 80¢. Looks like a good deal.

Notice the units of the Lagrange Multiplier in the previous problem. Maximum profit $\bar{P}$ is in dollars, and L is in hours (of labor). Thus, $d\bar{P}/dL$ has units dollars per hour. Those units are a *price* – the price of an additional hour of labor in this problem. Because the units of the Lagrange Multiplier are price, the multiplier is called a *shadow price*. Some people use the term *shadow price* for every Lagrange Multiplier because of the prevalence of problems in which we are thinking about the cost of changing some parameter. Notice that in the language of marginals, the shadow price in the present problem is the marginal maximum profit of labor hours.

You will do several problems that use Lagrange Multipliers to evaluate the cost of resources to make decisions concerning them.

CHAPTER 7

The Integral

1. Antiderivatives

Recall that a function that has a derivative is called *differentiable*. All functions that have algebraic formulas, or that involve the exponential function or the natural logarithm, are differentiable on every open interval where they are defined. Not all functions are differentiable – this is a difficult and technical question that we will not address in general.

We have worked extensively with the problem of finding the derivative $f'(x)$ for given functions $f(x)$. Now we want to reverse that process, for there are many situations in which the *derivative* is given, and we want to find the function. For instance, we might know the velocity $v(t)$ of an object moving on the x-axis and be seeking the position $x(t)$. The notation $g(x) = f'(x)$ says that $g(x)$ is the derivative of $f(x)$; to reverse the description we say that $f(x)$ is an *antiderivative* of $g(x)$. For instance, since

$$\left[3x^2 - \ln(x)\right]' = 6x - \frac{1}{x}$$

we have that $3x^2 - \ln(x)$ is an antiderivative of $6x - 1/x$.

The derivative of a constant is 0. So, if we add a constant k to the anti-derivative:

$$3x^2 - \ln(x) + k$$

then we get the same derivative

$$\left[3x^2 - \ln(x) + k\right]' = 6x - \frac{1}{x} + 0$$

Taking $k = 6$ as a specific example, **both** $3x^2 - \ln(x)$ and $3x^2 - \ln(x) + 6$ are antiderivatives of $6x - 1/x$.

It turns out that *all* the antiderivatives of a given function can be described in this way: find a particular antiderivative, and the rest of them are obtained by adding constants.[1] Thus, since $(x^3)' = 3x^2$, we have

$$\text{If} \quad f'(x) = 3x^2 \quad \text{then} \quad f(x) = x^3 + k \quad \text{for some constant} \quad k$$

We introduce the most common notation for the antiderivative. We will concentrate on using this notation; it will be explained more fully in the next section. We write

$$\int f(x) \cdot dx$$

for an antiderivative of $f(x)$, also called the *indefinite integral* of $f(x)$. The $\int$ is the *integral sign*; it originated as a German S, standing for *summe*, recognizable as the the German word for *sum*. The function $f(x)$ is the *integrand*, and the differential dx indicates the variable. Since $(x^2)' = 2 \cdot x$, we write

(7.1) $$x^2 = \int 2 \cdot x \cdot dx$$

We have $(x^2 + 3)' = 2 \cdot x$, and so

(7.2) $$x^2 + 3 = \int 2 \cdot x \cdot dx$$

As we saw before, different functions can have the same derivative; that's why the indefinite integral is indefinite! We will see that ambiguity is an advantage in this setting. Some texts write

$$\int 2 \cdot x \cdot dx = x^2 + k \quad \text{where } k \text{ is a constant}$$

This point of view regards the indefinite integral as standing for *all* antiderivatives of $2 \cdot x$ simultaneously. Our preference is to use the indefinite integral

---

[1]This fact is a consequence of the Mean Value Theorem. You might look up a statement of that result in a standard calculus text. When the Mean Value Theorem is expressed in terms of the secant line on a graph, it seems obvious to most people.

to stand for *some antiderivative* rather than *all of them.* The thing to remember is that if you have one antiderivative of a function, you can get the rest of them by adding constants.

Each rule for calculating derivatives can be turned into a rule for calculating antiderivatives. We will begin with the power rule.

Inverse Power Rule   Let n be a number with $n \neq -1$. Then

$$\int x^n \cdot dx = \frac{x^{n+1}}{n+1}$$

Explanation.   We use the Power Rule for derivatives to compute

$$\left[\frac{x^{n+1}}{n+1}\right]' = (n+1) \cdot \frac{x^n}{n+1} = x^n$$

and this proves that our Inverse Power Rule is correct. ■

We disallow $n = -1$ in the Inverse Power Rule for the simple reason that when $n = -1$, the $n + 1$ in the denominator would be 0. The case $n = -1$ has its own formula. Remember that $x^{-1} = 1/x$.

Logarithm Rule   We have

$$\int \frac{dx}{x} = \ln|x| \quad \text{when} \quad x \neq 0$$

Explanation.   Remember that $\ln(a)$ is defined only when $a > 0$. When $x > 0$, we know that

$$\ln'(x) = \frac{1}{x} \quad \text{and so} \quad \ln(x) = \int \frac{dx}{x} \quad \text{when} \quad x > 0$$

We will need to have an antiderivative for $1/x$ in the case that $x < 0$, as well. In that case, notice that $-x > 0$, and so $\ln(-x)$ is defined. Use the Chain Rule to compute:

$$\left[\ln(-x)\right]' = \ln'(-x) \cdot (-x)' = \frac{1}{-x} \cdot (-1) = \frac{1}{x}$$

Thus,

$$\ln(-x) = \int \frac{dx}{x} \quad \text{when} \quad x < 0$$

When $-x < 0$, we can write $-x = |x|$; when $x > 0$, we have $x = |x|$, so we have the Logarithm Rule in both cases. ■

Don't forget the absolute value in $\ln|x|$. We will need to allow x to be negative, and the logarithm of a negative is undefined, but $|x|$ is always positive if $x \neq 0$, and so $\ln|x|$ is defined.

The exponential function is its own derivative, and so it is its own antiderivative.

Exponential Rule   We have

$$\int e^x \cdot dx = e^x$$

■

Recall the Addition Rule and Constant Multiple Rule for derivatives:

$$\left[f(x) + g(x)\right]' = f'(x) + g'(x) \quad \text{and} \quad \left[c \cdot f(x)\right]' = c \cdot f'(x)$$

where $f(x)$ and $g(x)$ are differentiable, and c is a constant. It follows that we have an Addition Rule and a Constant Multiple Rule for the indefinite integral.

Addition Rule, Constant Multiple Rule   If $f(x)$ and $g(x)$ are differentiable, and if c is a constant, then

$$(7.3) \qquad \int \left[f(x) + g(x)\right] \cdot dx = \int f(x) \cdot dx + \int g(x) \cdot dx$$

and

$$(7.4) \qquad \int c \cdot f(x) \cdot dx = c \cdot \int f(x) \cdot dx$$

Here is a sample use.

Example. We compute

$$\int \left[2 \cdot e^x + \frac{5}{x^3}\right] \cdot dx = \int 2 \cdot e^x \cdot dx + \int \frac{5}{x^3} \cdot dx \qquad \text{addition rule}$$

$$= 2 \cdot \int e^x \cdot dx + 5 \cdot \int x^{-3} \cdot dx \qquad \text{constant multiple}$$

$$= 2 \cdot \cdot e^x + 5 \cdot \frac{x^{-2}}{-2} \qquad \text{exp, inverse power}$$

$$= 2 \cdot e^x - \frac{5}{2x^2}$$

∎

In the following problem, pay attention to the use of the unknown constant K.

Problem T7.1. A steel bar 2 feet long has density $10x^2 - x^3 + 2e^x$ pounds per foot at the point x feet from one end of the bar. What is the weight of the bar?

Solution. Measure density and weight $W(x)$ from the left hand end of bar. Then $W'(x)$ is the density, so that $W'(x) = 10x^2 - x^3 + 2e^x$, and so

$$W(x) = \int \left[10x^2 - x^3 + 2e^x\right] \cdot dx$$

Let's use the various rules on the right side, and then we'll bring $W(x)$ back into the picture. Make sure you understand which rules are being used in each line of the following.

$$\int \left[10x^2 - x^3 + 2e^x\right] \cdot dx$$

$$= \int 10x^2 \cdot dx - \int x^3 \cdot dx + \int 2e^x \cdot dx$$

$$= 10 \cdot \int x^2 \cdot dx - \int x^3 \cdot dx + 2 \cdot \int e^x \cdot dx$$

$$= 10 \cdot \frac{x^3}{3} - \frac{x^4}{4} + 2 \cdot e^x = \frac{10}{3}x^3 - \frac{1}{4}x^4 + 2e^x$$

The function we ended up with is an antiderivative of the density, and $W(x)$ is an antiderivative of the density. It follows that

$$(7.5) \qquad\qquad W(x) = \frac{10}{3}x^3 - \frac{1}{4}x^4 + 2e^x + K$$

for some constant K. We can figure out what K is by plugging in any particular value of x. Go back to the definition of $W(x)$ to see that $W(0) = 0$. Let $x = 0$ in (7.5) and we have

$$0 = 2 + K \quad \text{so that} \quad -2 = K$$

This shows that

$$W(x) = \frac{10}{3}x^3 - \frac{1}{4}x^4 + 2e^x - 2$$

Finally, the weight of the bar is $W(2)$:

$$W(2) = \frac{10}{3} \cdot 8 - \frac{1}{4} \cdot 16 + 2e^2 - 2 \approx 35.44 \text{ pounds}$$

■

In the previous problem, we used the antiderivative rules to find the indefinite integral of the density. The weight function is a *definite* antiderivative. The constant K synchronizes the indefinite antiderivative to the definite one of the problem.

Sometimes we are asked to verify an antiderivative formula: that's just asking us to take the derivative.

Problem T7.2. Show that

$$x + 7 \cdot \ln|x - 7| = \int \frac{x}{x - 7} \cdot dx \quad \text{for} \quad x \neq 7$$

Solution. As we said, we are just being asked to take the derivative; we need to remember that $(\ln|x|)' = 1/x$ no matter whether $x > 0$ and $x < 0$.

Compute

$$\big[x + 7 \cdot \ln|x - 7|\big]' = 1 + \frac{7}{x - 7}$$

We get the integrand with a little algebra.

$$1 + \frac{7}{x-7} = \frac{x-7}{x-7} + \frac{7}{x-7} = \frac{x}{x-7}$$

as needed. ■

There is an antiderivative rule that corresponds to the Chain Rule. While we will not pursue that correspondence in general, we will use the following special case.

Linear Inside Rule   Suppose that a, b are constants with $a \neq 0$, and suppose that $f'(x)$ is differentiable. Then

$$\int f'(ax+b) \cdot dx = \frac{1}{a} \cdot f(ax+b)$$

Explanation   Take the derivative of the right side, using the Chain Rule.

$$\left[\frac{1}{a} \cdot f(ax+b)\right]' = \frac{1}{a} \cdot f'(ax+b) \cdot (ax+b)' = \frac{1}{a} \cdot f'(ax+b) \cdot a = f'(ax+b)$$

■

The Linear Inside Rule is somewhat abstract. If we ignore the linear term $ax+b$, we see the pattern $\int f'(x) \cdot dx = f(x)$. The Rule says to imitate this pattern, carrying the $ax+b$ along the way we do in the Chain Rule. But we modify the antiderivative with the fraction $1/a$, dividing by the slope of the linear term $ax+b$. Let's see some uses of this rule.

Problem T7.3. Compute each of the following.

   (a) $\displaystyle\int (3x-2)^4 \cdot dx$ **(b)** $\displaystyle\int \exp(-2t) \cdot dt$ **(c)** $\displaystyle\int \sqrt{3y+1} \cdot dy$

Solution.   For (a), we use that

$$\int x^4 \cdot dx = \frac{x^5}{5}$$

In the notation of the Linear Inside Rule, $f'(x) = x^4$. Notice that the integral in (a) involves the fourth power function with $3x-2$ inside – that's the linear inside, the $ax + b$ expression. We compute the antiderivative of $(3x - 2)^4$ by using the antiderivative of the fourth power, dividing by the slope $a = 3$.

$$\int (3x - 2)^4 \cdot dx = \frac{1}{3} \cdot \frac{(3x - 2)^5}{5}$$

For (b), the linear inside on $\exp(-2t)$ is $-2t$, which has slope $a = -2$. Thus,

$$\int \exp(-2t) \cdot dt = \frac{1}{-2} \cdot \exp(-2t)$$

For (c), we treat the square root the way we did the fourth power in (a):

$$\int \sqrt{3y + 1} \cdot dy = \int (3y + 1)^{1/2} \cdot dy = \frac{1}{3} \cdot \frac{(3y + 1)^{3/2}}{3/2} = \frac{2}{9} \cdot (3y + 1)^{3/2}$$

For many integrals, there is more than one way to get the answer. For instance, consider

$$\int (5 - 2 \cdot t) \cdot dt$$

The most direct approach is to use the rules for polynomials:

$$\int (5 - 3 \cdot t) \cdot dt = 5 \cdot t - 2 \cdot \frac{t^2}{2} = 5 \cdot t - t^2$$

There is another way! We see that $(5 - 2 \cdot t)^1$ is linear inside the first power, and so, remembering to divide by the slope

$$\int (5 - 3 \cdot t) \cdot dt = \frac{1}{-2} \frac{(5 - 2 \cdot t)^2}{2} = -\frac{1}{4} \cdot (5 - 2 \cdot t)^2$$

It is instructive to compare the two answers. Multiplying out the second one:

$$-\frac{1}{4} \cdot (5 - 2 \cdot t)^2 = -\frac{1}{4} \cdot (25 - 20 \cdot t + 4 \cdot t^2) = -\frac{25}{4} + 5 \cdot t - t^2$$

This is the first answer, with a constant added. We can always add a constant to an antiderivative to get another one.

2. The Definite Integral

We introduce the *definite integral*, one of the most important and commonly occurring mathematical objects in all of science. We will discuss the definition and computation of the integral, and we will give the two main ways it occurs in applications.

There are many approaches to this material. We want to start with the computation of the definite integral. That approach can be presented by building on the indefinite integral studied in the previous section.

First, we remind you about the *closed interval* notation. For instance, $[1, 4]$ stands for the set of all numbers x with $1 \le x \le 4$. In general, if a, b are numbers with $a \le b$, the closed interval $[a, b]$ is the set of all numbers x with $a \le x \le b$.

Second, we introduce notation that will help us evaluate functions at various values. Given $F(x)$ and numbers a, b, we define

$$F(x)\Big|_a^b = F(b) - F(a)$$

Plug in the top number first, then subtract what you get when you plug in the bottom number. Thus,

$$x^2\Big|_2^3 = 3^2 - 2^2 = 5 \quad \text{and} \quad \exp(x/2)\Big|_0^2 = \exp(1) - \exp(0) = e - 1$$

You might recognize that $F(x)\Big|_a^b$ is ΔF. Hold that thought.

You have seen the *indefinite integral*

$$\int f(x) \cdot dx$$

as a notation for an antiderivative of $f(x)$. If $f(x)$ is defined for every value of x on a closed interval $[a, b]$, the *definite integral of $f(x)$ over $[a, b]$* is written

$$\int_a^b f(x) \cdot dx$$

As in the indefinite integral, the function $f(x)$ is called the *integrand* and the differential dx indicates the variable x for which $f(x)$ is a function. The number a is the *lower limit of integration*, and the number b is the *upper limit of integration*.

We have many things to say about the definite integral; we begin with how it is computed in most cases. First, find an antiderivative[2] $F(x)$ for $f(x)$ on $[a, b]$. Then,

$$\int_a^b f(x) \cdot dx = F(x)\Big|_a^b = F(b) - F(a)$$

For instance, the Inverse Power Rule says that

(7.6)
$$\int x^2 \cdot dx = \frac{1}{3} \cdot x^3$$

And we compute the definite integral:

$$\int_2^4 x^2 \cdot dx = \frac{1}{3} \cdot x^3\Big|_2^4 = \frac{1}{3} \cdot 4^3 - \frac{1}{3} \cdot 2^3 = \frac{56}{3}$$

We know that x^2 has other antiderivatives. For instance,

$$\frac{1}{3} \cdot x^3 + 5 = \int x^2 \cdot dx$$

What if we had used this antiderivative rather than just $x^3/3$?

$$\int_2^4 x^2 \cdot dx = \left(\frac{1}{3} \cdot x^3 + 5\right)\Big|_2^4$$

$$= \left(\frac{1}{3} \cdot 4^3 + 5\right) - \left(\frac{1}{3} \cdot 2^3 + 5\right)$$

$$= \frac{64}{3} + 5 - \frac{8}{3} - 5 = \frac{56}{3}$$

---

[2]To get very technical, the mere existence of an antiderivative for $f(x)$ is not enough to ensure that the definite integral exists. If the function $f(x)$ is differentiable, however, then the definite integral will exist, as will an antiderivative for $f(x)$, and our linking of the antiderivative and the definite integral will be correct. All of the functions we will consider are differentiable.

We see that the constant 5 cancels! This is what happens in general, for suppose we have an antiderivative $F(x)$ for $f(x)$, so that $F'(x) = f(x)$. In Section 1 we said that any other antiderivative for $f(x)$ is of the form $F(x)+K$ where K is a constant. Using $F(x) + K$ in place of $F(x)$ in the definition of the definite integral we get

$$\int_a^b f(x) \cdot dx = (F(x) + K)\Big|_a^b$$
$$= (F(b) + K) - (F(a) + K)$$
$$= F(b) + K - F(a) - K = F(b) - F(a)$$

Thus, we compute the same value for the definite integral no matter which antiderivative is used.

We see that the value of the definite integral is ΔF, where $F'(x) = f(x)$ is the integrand. That leads to many applications. Here is an example.

Problem T7.4. An object is moving along the x-axis with velocity $v(t) = 5 - 2 \cdot t^2$, for $t \geq 0$. What is the displacement as t goes from 1 to 3?

Solution. Recall that displacement is Δx where x is the position. Write $x(t)$ to remind us that x is a function of t. Since t goes from 1 to 3, we are asking for

$$\Delta x = x(3) - x(1)$$

We know that velocity $v(t) = x'(t)$, and so $x(t)$ is an antiderivative of $v(t)$. Thus,

$$x(3) - x(1) = x(t)\Big|_1^3 = \int_1^3 v(t) \cdot dt = \int_1^3 (5 - 2 \cdot t^2) \cdot dt$$
$$= 5 \cdot t - 2 \cdot \frac{t^3}{3}\Big|_1^3 = \left[5 \cdot 3 - \frac{2}{3} \cdot 3^3\right] - \left[5 \cdot 1 - \frac{2}{3}\right] = -\frac{22}{3}$$

■

There is a subtlety in the solution to the previous problem. Our calculation of the definite integral used the antiderivative

(7.7) $$5 \cdot t - 2 \cdot \frac{t^3}{3} = \int (5 - 2 \cdot t^2) \cdot dt$$

The integrand is velocity, and $x(t)$ is an antiderivative of velocity. Does that mean that x is the left side of (7.7)? Not necessarily! It would be the case that

$$x(t) = 5 \cdot t - 2 \cdot \frac{t^3}{3} + K$$

for some constant K, but the constant does not affect the calculation of the definite integral. The antiderivative we used and the antiderivative $x(t)$ give the same answer.

Here are some basic properties of the definite integral. Throughout, each integrand is assumed to be defined and differentiable between the limits of integration. The first two properties say that the integral is *linear*, that integration distributes through sums and across constants. These identities follow directly from the corresponding identities (7.3) and (7.4) for indefinite integrals, as we will point out in class. The third property shows how to chain integrals together. These properties are fairly natural, and we will usually use them without calling attention.

(7.8) $$\int_a^b (f(x) + g(x)) \cdot dx = \int_a^b f(x) \cdot dx + \int_a^b g(x) \cdot dx$$

(7.9) $$\int_a^b C \cdot f(x) \cdot dx = C \cdot \int_a^b f(x) \cdot dx$$
$$\text{where } C \text{ is constant}$$

(7.10) $$\int_a^b f(x) \cdot dx + \int_b^c f(x) \cdot dx = \int_a^c f(x) \cdot dx$$

We will verify (7.10); let $F'(x) = f(x)$.

$$\int_a^b f(x) \cdot dx + \int_b^c f(x) \cdot dx = F(x)\Big|_a^b + F(x)\Big|_b^c$$
$$= F(b) - F(a) + F(c) - F(b)$$
$$= F(c) - F(a) = \int_a^c f(x) \cdot dx$$

and (7.10) holds.

3. Riemann Sums

We have computed the definite integral from an antiderivative for the integrand. There are many functions for which there is no obvious formula for the antiderivative – for instance, integrals involving the function $\exp(-x^2)$ come up often in statistics. How are we to compute the definite integral when we can't find an antiderivative? If you accept the statement we made to the effect that the integral occurs often, you might find this question compelling.

There is another problem with our approach to the integral. Many applications of the integral do not involve an antiderivative in any obvious way.[3] The link from those applications to the integral involves looking at the integral differently. It turns out that this alternative point of view answers the question posed in the last paragraph about the existence of antiderivatives.

Let's introduce this new idea via a very specific example. We need a function defined on a closed interval; let's use $x^2 - 3$ on $[1, 3]$. Now we divide the interval $[1, 3]$ into some number of equal subintervals. Let's divide it into 5 subintervals; the length of $[1, 3]$ is $3 - 1 = 2$, and so each of the 5 pieces will have length $2/5 = 0.4$. Here are the subintervals.

$$\begin{array}{cccccc} 1 & 1.4 & 1.8 & 2.2 & 2.6 & 3 \end{array}$$

---

[3]For instance, see the application example on p.139 of Chapter 8.

Next we add up the function values at the right hand endpoint of each subinterval. Our function was $x^2 - 3$, and here are the values:

$$\begin{array}{cccccc} x: & 1.4 & 1.8 & 2.2 & 2.6 & 3 \\ x^2 - 3: & -1.04 & 0.24 & 1.84 & 3.76 & 6 \end{array}$$

And their sum:

$$-1.04 + 0.24 + 1.84 + 3.76 + 6 = 10.8$$

Finally, we multiply by the width of the subintervals; in this case, the width is 0.4. The product of the sum and width is called a *Riemann sum*.

$$10.8 \cdot 0.4 = 4.32$$

A single Riemann sum is not impressive, and the calculation of specific examples dissolves into arithmetic. What is interesting is what happens as the number of intervals used gets larger and larger. Here are the Riemann sums for $x^2 - 3$ on $[1, 3]$ when we divide up the interval into n pieces for various n. (In class we will explain how the sums were computed; the answers have been rounded.)

$$\begin{array}{cccccc} n: & 5 & 10 & 100 & 1000 & 10^6 \\ \text{Riemann sum}: & 4.32 & 3.48 & 2.75 & 2.675 & 2.667 \end{array}$$

Our table shows only a few sums, but as the number n of subintervals gets larger and larger, the Riemann sums get closer and closer to (about) 2.667. That number turns out to be the value of the definite integral using the same function $x^2 - 3$ and the same interval $[1, 3]$:

$$\int_1^3 (x^2 - 3) \cdot dx = \left. \frac{x^3}{3} - 3 \cdot x \right|_1^3 = 0 - \left(-\frac{8}{3} \right) = \frac{8}{3} \approx 2.67$$

It turns out that Riemann sums can be made close to the definite integral by using more and more subintervals.

Let's describe what we just did more generally. We are given a function $f(x)$ defined on a closed interval $[a, b]$. Given a positive integer n, there is a

Riemann sum which we will denote[4] $\mathcal{R}(f(x), a, b, n)$. The sum is formed by dividing the interval $[a, b]$ into n equal subintervals, each of width $(b - a)/n$. Denote the endpoints of the subintervals like this.

$$(7.11) \qquad a = x_0 < x_1 < x_2 < \cdots < x_{n-1} < x_n = b$$

We write $a = x_0$ and $b = x_n$ to make the notation for the subintervals more uniform.[5] Here are the subintervals:

$$[x_0, x_1], \ [x_1, x_2], \ [x_2, x_3], \ \cdots \ [x_{n-1}, x_n]$$

Here is the Riemann sum:

$$(7.12) \qquad \mathcal{R}(f(x), a, b, n) = \frac{b - a}{n} \cdot \Big[f(x_1) + f(x_2) + f(x_3) + \cdots + f(x_n) \Big]$$

If $f(x)$ is differentiable, then it turns out that as n gets larger and larger, the sums approach a specific number; this number is the *definite integral*. In notation

$$(7.13) \qquad \lim_{n \to \infty} \mathcal{R}(f(x), a, b, n) = \int_a^b f(x) \cdot dx$$

The notation $n \to \infty$ simply means that n gets larger and larger, without bound. We are saying that the larger n gets, the closer the Riemann sums get to the definite integral.

In our Riemann sums, the terms $f(x_j)$ evaluate the function $f(x)$ at the right endpoint of each of the subintervals into which the main interval is partitioned. It is just as acceptable to use the left endpoints, or, in fact, any point in each of the subintervals. The points can be chosen systematically or randomly, as long as there is a term $f(c_j)$ for some point c_j in the j-th interval. Computer programs that calculate Riemann sums numerically use a variety of schemes. And those schemes are important, since one use of (7.13) is to give

---

[4]There is no universally used notation for Riemann sums. The notation we introduce will be used only in this course.

[5]We won't need a precise formula for the x_j, although, in case you are interested, $x_j = a + j \cdot (b - a)/n$.

numerical approximations to definite integrals. We will provide a web function for that purpose.

Problem T7.5. Demonstrate (7.13) when $f(x) = 2 \cdot x + 5$ on $[0, 6]$.

Solution. We need to pick various n and form the Riemann sum (7.12). The problem doesn't tell us how to do this. We'll start with $n = 8$ and compute the sum by hand. When the interval $[0, 6]$ is divided into 8 subintervals, each has width $6/8 = 0.75$. Here are the endpoints, as in (7.11).

$$0 < 0.75 < 1.5 < 2.25 < 3 < 3.75 < 4.5 < 5.25 < 6$$

The Riemann sum (7.12) involves these function values:

x	: 0.75	1.5	2.25	3	3.75	4.5	5.25	6
$2 \cdot x + 5$:	6.5	8	9.5	11	12.5	14	15.5	17

We need to add up the function values and multiply by the width 0.75.

$$\mathcal{R}(2x + 5, 0, 6, 8) = 0.75 \cdot \left[6.5 + 8 + 9.5 + 11 + 12.5 + 14 + 15.5 + 17\right]$$
$$= 70.5$$

We computed the sums for other values of n numerically.

n :	100	1000	5000	10000	50000
Riemann sum :	66.36	66.036	66.0072	66.0036	66.00072

Numerical evidence will never be absolutely convincing, but it looks as if the Riemann sums are getting close to 66.

We know how to calculate the definite integral:

$$\int_0^6 (2 \cdot x + 5) \cdot dx = x^2 + 5 \cdot x \Big|_0^6 = 66 - 0 = 66$$

∎

Here is an integral for which the approximation is essential, since we don't know how to find an antiderivative for the integrand.

Problem T7.6. Approximate the following integral.

$$\int_{-2}^{2} x^2 \cdot \exp(-x^2) \cdot dx$$

Solution. We computed Riemann sums for the following values of n.

n :	100	1000	5000
Riemann Sum :	0.84539	0.84545	0.84545

We are pretty confident that we have 5 decimal places for the integral. ■

There is an important theoretical point here. Riemann sums can be used to approximate the integral from the integrand – you don't need to know an antiderivative. It turns out that Riemann sums can be used to *define* an antiderivative for a given differentiable function $f(x)$. Indeed, we fix a left endpoint a, and then, for each $t \geq a$, we define

$$F(t) = \int_{a}^{t} f(x) \cdot dx$$

where the integral is defined by Riemann sums, using (7.13). It can be shown that $F'(x) = f(x)$; this is that $F(x)$ is an antiderivative for $f(x)$. The point is that a function that is differentiable on an interval always has an antiderivative there, whether we can find a nice formula for that antiderivative, or not.

The link between antiderivatives and Riemann sums is called the *Fundamental Theorem of Calculus*. This theorem summarizes our work in this chapter.

FUNDAMENTAL THEOREM OF CALCULUS. *Let $f(x)$ be a differentiable function on $[a, b]$. Then the equation (7.13) holds. Furthermore, $f(x)$ has an antiderivative $F(x)$, defined for all x on the interval $[a, b]$, and*

$$\int_{a}^{b} f(x) \cdot dx = F(b) - F(a)$$

■

CHAPTER 8

Interpreting and Using the Integral

1. Anti-Rates

If y is a function of time t, then the *rate of change* of y is the derivative dy/dt. If we are given the rate of change $g(t)$ of y, then

$$y = \int g(t) \cdot dt$$

We lightheartedly refer to y as the *anti-rate* of $g(t)$. Also remember that Δy is a definite integral. Both these ideas nurture applications.

Problem T8.1. A population grows at rate $10 \cdot (10 - t)$ individuals per year, where t is time in years. Assuming that the population starts at 0, what is the population after 15 years? When does the population reach 0?

Solution. For the first question, we notice that if the population is $P(t)$, then the rate of growth is $P'(t)$. Thus,

$$P'(t) = 10 \cdot (10 - t)$$

The population after 15 years is $P(15)$. Since $P(0) = 0$ (population starts at 0), we see that

$$P(15) = P(15) - P(0) = \int_0^{15} P'(t) \cdot dt$$

And we compute

$$\int_0^{15} P'(t) \cdot dt = \int_0^{15} 10 \cdot (10 - t) \cdot dt = 10 \cdot \left[10 \cdot t - \frac{t^2}{2} \right] \Big|_0^{15} = 375$$

Thus, $P(15) = 375$.

To see when the population is 0, we solve for the time $t = c$.

$$0 = 10 \cdot \int_0^c (10 - t) \cdot dt = 10 \cdot \left[10 \cdot t - \frac{t^2}{2} \right] \Big|_0^c = 10 \cdot \left[10 \cdot c - \frac{c^2}{2} \right]$$

Dividing by 10 and rearranging, we have $10c = c^2/2$. We see that $c = 0$ is one solution; not surprising, since we start with population 0. If $c \neq 0$, we can divide by c, and get

$$10 = \frac{c}{2} \quad \text{so that} \quad c = 20$$

After 20 years, the population has died off. ∎

To compute $P(15)$ in the previous problem, we used a definite integral – that worked because we could write

$$P(15) = P(15) - P(0) = \Delta P$$

Let's sketch an alternative way to do the same problem: we have $P'(t) = 10 \cdot (10 - t)$, and so we can write $P(t)$ as an indefinite integral:

$$P(t) = \int 10 \cdot (10 - t) \cdot dt = 10 \cdot \left[10 \cdot t - \frac{t^2}{2} \right] + K = 100 \cdot t - 50 \cdot t^2 + K$$

We include the constant K to synchronize the antiderivative we found from the power rule with P – a specific anti-derivative. Since we know that $P(0) = 0$, we can compute K by plugging in $t = 0$. We get $0 = P(0) = K$. Now we have the more specific formula $P(t) = 100 \cdot t + 50 \cdot t$. The problem asks for $P(15)$, and we get that by plugging in $t = 15$.

Problem T8.2. An *income stream* is defined by a rate at which money comes in. Say we receive $1000 \exp(-0.03t)$ \$/year for 7 years.[1] What is the total amount of money received?

Solution. If $A(t)$ is the amount of money received up to time t, then $A'(t)$ is the rate at which money is received. In other words, $A'(t) = 1000 \exp(-0.03t)$.

---

[1] Notice that the rate is not constant during a year. Thus, this model does not imagine that we are paid each year, but continually as time goes along. That's why the units \$ per year indicate a *derivative*.

Also, $A(0) = 0$, and the total amount received is $A(7)$. Thus,

$$A(7) = A(7) - A(0) = \int_0^7 A'(t) \cdot dt$$

$$= \int_0^7 1000 \cdot \exp(-0.03t) \cdot dt = \frac{1000}{-0.03} \cdot \exp(-0.03t)\Big|_0^7$$

$$= -\frac{1000 \cdot \exp(-0.03 \cdot 7)}{0.03} + \frac{1000 \cdot \exp(0)}{0.03} \approx \$6313.86$$

■

As with the population problem, we can do the income stream problem with an indefinite integral instead of a definite one. We may discuss this in class.

Problem T8.3. An object moving along the y-axis has acceleration $3\sqrt{t}$. The initial velocity of the object is -2. What is its displacement over $1 \le t \le 4$?

Solution. If v is the velocity, then dv/dt is the acceleration. Thus

$$v = \int 3\sqrt{t} \cdot dt = 3 \cdot \frac{t^{3/2}}{3/2} + C = 2 \cdot t^{3/2} + C$$

where C is a constant. When $t = 0$, we have $v = -2$ (the initial velocity), and so $C = -2$.

Writing $y(t)$ to indicate that y is a function of t, we have $y'(t) = v$. The problem asks for the displacement of y when $1 \le t \le 4$. That displacement is

$$\Delta y = y(4) - y(1) = \int_1^4 y'(t) \cdot dt$$

$$= \int_1^4 v \cdot dt = \int_1^4 \left[2t^{3/2} - 2 \right] \cdot dt$$

$$= 2 \cdot \frac{t^{5/2}}{5/2} - 2t \Big|_1^4 = \frac{4}{5} \cdot t^{5/2} - 2t \Big|_1^4$$

$$= \left(\frac{4}{5} \cdot 4^{5/2} - 8 \right) - \left(\frac{4}{5} - 2 \right) = \frac{94}{5}$$

■

Problem T8.4. We have an empty cylindrical container of radius 2 inches. Water is poured in at a rate of $10 - t$ cubic inches per second, for $0 \leq t \leq 10$. What is the height of the water at the end?

Solution. The rate in cubic inches per second is the derivative dV/dt where V is the volume. Thus, V is an antiderivative of $10 - t$. Also, $V(0) = 0$, since the container starts empty. At time $t = 10$, we have

$$V(10) = V(10) - V(0) = \int_0^{10} V'(t) \cdot dt = \int_0^{10} (10 - t) \cdot dt$$

$$= 10t - \frac{t^2}{2} \Big|_0^{10} = 100 - \frac{100}{2} = 50 \text{ cubic inches}$$

The volume of a cylinder satisfies $V = \pi \cdot r^2 \cdot h$, where r is the radius and h is the height. We have $r = 2$ and $V = 50$, and we can solve for h:

$$h = \frac{V}{\pi \cdot r^2} = \frac{50}{\pi \cdot 4} \approx 3.98 \text{ inches}$$

■

There is another way to do the previous problem; we can focus on height h from the get-go. We have $V = \pi \cdot r^2 \cdot h$; since $r = 2$, this is $V = 4 \cdot \pi \cdot h$. Taking the derivative with respect to time t:

$$\frac{dV}{dt} = 4 \cdot \pi \cdot \frac{dh}{dt}$$

The problem tells us dV/dt:

$$10 - t = 4 \cdot \pi \cdot \frac{dh}{dt} \quad \text{so that} \quad \frac{10 - t}{4 \cdot \pi} = \frac{dh}{dt}$$

We can find h with an antiderivative, just as we found V.

2. Area

Let $f(x)$ be a differentiable and non-negative function on the interval $[a, b]$. Let R be the plane figure[2] bounded by the x-axis, the curve $y = f(x)$, the line $x = a$, and the line $x = b$. Here is a generic picture of R.

---

[2]A *plane figure* is nothing more than a set of points in the plane.

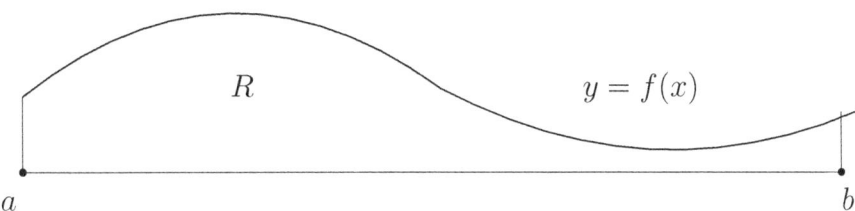

It turns out that

(8.1) $$\text{area of } R = \int_a^b f(x) \cdot dx$$

We want to sketch an explanation for this very important formula.

During the early 1600's Fermat and others were able to calculate the areas of various plane figures, particularly when the boundary of the figure is a polynomial function. It was noticed that the formulas obtained seemed to involve antiderivatives of the boundary functions. See [**11**, Chapter 4]. Newton's teacher Barrow drew up lectures on this subject – lectures that he turned over to Newton to complete. Newton developed a much more comprehensive and succinct theory, and he was able to sketch out a convincing argument for (8.1). Newton's idea was a key factor in the development of the general calculus. Leibnitz developed similar ideas at about the same time as Newton, and the integral notation we use goes back to Leibnitz. See [**11**, Chapter 5].

We choose a number z between a and b, and let $A(z)$ be the area between $y = f(x)$ and the x-axis and between $x = a$ and $x = z$. Here is a generic picture.

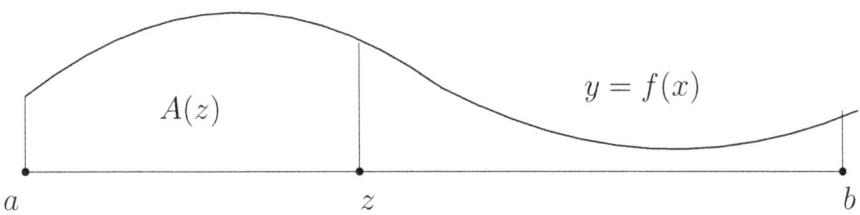

What Newton realized is that ΔA is itself an area. Here is a picture in the case that $\Delta z > 0$.

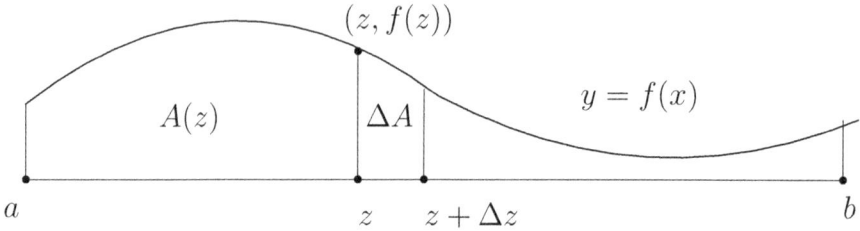

Imagining a very thin such area, ΔA is almost a rectangle with base Δz and height $f(z)$, for $f(z)$ is the height of the left-hand vertical side of ΔA. In other words,

$$\Delta A \approx f(z) \cdot \Delta z \quad \text{which is} \quad \frac{\Delta A}{\Delta z} \approx f(z)$$

Newton assumed that if we let $\Delta z \to 0$ the approximation becomes an equation. More modern mathematics can be used to show that this is, in fact, the case. Thus, we get

(8.2) $A'(z) = f(z)$

The derivative of the area function is the boundary function. Notice how the variable x in $f(x)$ changes to a z when we take the derivative of the area function. The variable z is a specific value of the variable x.

Let us see how the foregoing leads to (8.1) in a special case.

Problem T8.5. Find the area bounded by the x-axis, the curve $y = 1/x$, for $1 \le x \le 3$.

Solution. Observe that $1/x$ is positive for x between 1 and 3. Let $A(z)$ be the area under $y = 1/x$, and $1 \le x \le z$. Notice that $A(1) = 0$, since that's the area of a line segment. Also notice that $A(3)$ is the area we want. Thus,

(8.3) $\text{area} = A(3) = A(3) - A(1) = \displaystyle\int_1^3 A'(x) \cdot dx$

The reason for involving the derivative of A is that (8.2) shows that

$$A'(z) = \frac{1}{z} \quad \text{and so} \quad A'(x) = \frac{1}{x}$$

changing variable from z to x. Thus, (8.3) shows that

$$\text{area} = \int_1^3 \frac{1}{x} \cdot dx$$

and that is the formula (8.1).

So much for theory; how about calculating the area! Now that we have our integral formula, we have only to compute the integral:

$$\text{area} = \int_1^3 \frac{1}{x} \cdot dx = \ln|x|\Big|_1^3 = \ln(3) - \ln(1) = \ln(3)$$

∎

In the solution just done, we emphasized the use of Newton's function $A(z)$. The point of that function was to obtain the equation (8.1) and calculate the area by integration. Going forward, we will use (8.1) without appealing to the apparatus of Newton's area function.

Because $\ln|x|$ is an antiderivative of $1/x$, we see that areas under $y = 1/x$ are logarithms. Perhaps that is an unexpected interpretation of the logarithm.

Problem T8.6. Write the area of a circle of radius r as an integral, and compute the integral numerically when $r = 1$.

Solution. The equation of a circle of radius r with center at $(0,0)$ is $x^2 + y^2 = r^2$. The top half of the circle is defined by solving for $y \geq 0$ to get $y = \sqrt{r^2 - x^2}$, for $-r \leq x \leq r$. The area between the top half of the circle and the x-axis (diameter) is

$$\int_{-r}^r \sqrt{r^2 - x^2} \cdot dx$$

This is the area of a semi-circle of radius r, and so the area of the circle is twice this integral.

Our Riemann sum calculator gives the area when $r = 1$ as approximately 3.1416. (That's supposed to be π.) ∎

We do not know how to find an antiderivative for $\sqrt{r^2 - x^2}$. You have been told that the area of a circle of radius r is $\pi \cdot r^2$. If we define π by the area of a circle with $r = 1$, then, as in the previous problem,

$$\pi = 2 \cdot \int_{-1}^{1} \sqrt{1 - x^2} \cdot dx$$

Let's show how this integral leads to the $\pi \cdot r^2$ formula. Let

$$R(x) = 2 \cdot \int \sqrt{1 - x^2} \cdot dx$$

Even though we don't have a formula for $R(x)$, we will use it to good effect. We have

$$(8.4) \qquad \pi = R(x)\Big|_{-1}^{1} = R(1) - R(-1)$$

For $r > 0$, use the Chain Rule to compute

$$\frac{d}{dx} R\left(\frac{x}{r}\right) = R'\left(\frac{x}{r}\right) \cdot \frac{1}{r} = \sqrt{1 - \frac{x^2}{r^2}} \cdot \frac{1}{r}$$

$$= 2 \cdot \sqrt{\frac{r^2 - x^2}{r^2}} \cdot \frac{1}{r} = 2 \cdot \frac{\sqrt{r^2 - x^2}}{r} \cdot \frac{1}{r} = \frac{2}{r^2} \cdot \sqrt{r^2 - x^2}$$

This shows us that

$$r^2 \cdot R\left(\frac{x}{r}\right) = 2 \cdot \int \sqrt{r^2 - x^2} \cdot dx$$

and so the area of a circle of radius r is this:

$$2 \cdot \int_{-r}^{r} \sqrt{r^2 - x^2} \cdot dx = r^2 \cdot R\left(\frac{x}{r}\right)\Big|_{-r}^{r}$$

$$= r^2 \cdot (R(1) - R(-1))$$

Looking at (8.4), we see that the area of the circle of radius r is $\pi \cdot r^2$.

The previous argument involves the abstract existence of an antiderivative – a consequence of the Fundamental Theorem of Calculus. We will not pursue this sort of argument very often – there will be one more use in this chapter

– but we need to say that this kind of calculation has some important uses in applied mathematics.

We computed the area between the graph of a function and the x-axis, assuming that the function was above the x-axis. More generally, we can compute the area between $y = f(x)$ and $y = g(x)$ for $a \leq x \leq b$, assuming that $f(x) \geq g(x)$.

Area between cuves If $f(x) \geq g(x)$ for $a \leq x \leq b$, then the area between $y = f(x)$ and $y = g(x)$ on this interval is

$$\int_a^b (f(x) - g(x)) \cdot dx$$

Explanation We can add a constant c to $g(x)$ so that $g(x) + c > 0$; that puts $y = g(x) + c$ above the x-axis. Since $f(x) \geq g(x)$, it follows that $y = f(x) + c$ is also above the x-axis. The area between $y = f(x)$ and $y = g(x)$ is the same as the area between $y = f(x) + c$ and $y = g(x) + c$, since the latter area is just the former area moved up in the plane.

If you draw a generic picture, you will see that the area between $y = f(x) + c$ and $y = g(x) + c$ is the difference between the area under $y = f(x) + c$ and the x-axis over the interval $[a, b]$ and the area under $y = g(x) + c$ and the x-axis over $[a, b]$. Thus, the area we are trying to compute is this:

$$\int_a^b (f(x) + c) \cdot dx - \int_a^b (g(c) + c) \cdot dx$$

$$= \int_a^b f(x) \cdot dx + \int_a^b c \cdot dx - \int_a^b g(x) \cdot dx - \int_a^b c \cdot dx$$

$$= \int_a^b f(x) \cdot dx - \int_a^b g(x) \cdot dx$$

$$= \int_a^b (f(x) - g(x)) \cdot dx$$

We see that the constant c disappears. ■

The area formula shows that we integrate the *upper curve minus the lower curve*. We usually determine which is which from the graphs. In the next few problems we deliberately do not draw the graphs in the text; we want to do that interactively in class, or have you do it for yourself.

Problem T8.7. Find the area bounded by $y = x^2$ and $y = 2 \cdot x + 8$.

Solution. It is not hard to see how a parabola and line intersect. The area bounded by these curves must lie between them, and because the parabola is concave up, we expect the line to be the upper curve. Let's see: to find the boundaries of the area, we need to figure out where they intersect. The equation $x^2 = 2 \cdot x + 8$ leads to $x^2 - 2 \cdot x + 8 = 0$, which is $(x + 2) \cdot (x - 4) = 0$. We see that the figure is bounded by the interval $[-2, 4]$.

It is not hard to figure out which is the upper curve from the sign diagram for the difference. Or we can plug in a number in the interval $[-2, 4]$ (say $x = 0$). One way or another we see that $y = 2 \cdot x + 8$ is above $y = x^2$. Our area is

$$\int_{-2}^{4} \left[(2 \cdot x + 8) - x^2 \right] \cdot dx = \int_{-2}^{4} \left[2 \cdot x + 8 - x^2 \right] \cdot dx$$

$$= x^2 + 8 \cdot x - \frac{1}{3} \cdot x^3 \Big|_{-2}^{4}$$

$$= \left[16 + 32 - \frac{64}{3} \right] - \left[4 - 16 + \frac{8}{3} \right]$$

$$= 36$$

■

Problem T8.8. Find the area bounded by $y = 2 \cdot x^2 + 2$ and $y = x^2 + 3 \cdot x$ and $x \le 3$.

Solution. This one is harder to see, because both curves are concave up parabolas. First, let's find the intersection between the curves:

$$2 \cdot x^2 + 2 = x^2 + 3 \cdot x \quad \text{leads to} \quad (x - 1) \cdot (x - 2) = 0$$

The curves intersect at $x = 1$ and $x = 2$, and so they trap area between them there. Over that interval, plugging in $x = 1.5$ (or looking at the sign pattern), we see that $y = x^2 + 3 \cdot x$ is the upper curve, and $y = 2 \cdot x^2 + 2$ is the lower curve. The area of the bubble between the curves is

$$\int_1^2 \left[(x^2 + 3 \cdot x) - (2 \cdot x^2 + 2) \right] \cdot dx = \int_1^2 \left[3 \cdot x - x^2 - 2 \right] \cdot dx = \frac{1}{6}$$

But there's more! The boundary $x = 3$ must also be taken into account. Between $x = 2$ and $x = 3$, the curves trap another bubble of area. In that range, the curve $y = 2 \cdot x^2 + 2$ is the upper curve. (How did we know that?) The additional area is this:

$$\int_2^3 \left[(2 \cdot x^2 + 2) - (x^2 + 3 \cdot x) \right] \cdot dx = \int_2^3 \left[x^2 + 2 - 3 \cdot x \right] \cdot dx = \frac{5}{6}$$

The total area is $(1/6) + (5/6) = 1$. ∎

Problem T8.9. Find the area bounded by $y = e^x$ and $y = 2 - x$ and $x \geq 0$.

Solution. From the graph, we see that $y = 2 - x$ is the upper curve and $y = e^x$ is the lower curve, until they meet somewhere to the right of the y-axis. To find the intersection point, we need to solve $e^x = 2 - x$. This equation does not have a nice, algebraic solution, and so we use Newton's Method to approximate the solution.[3] The equation is $e^x + x - 2 = 0$. Starting with $x = 0$, we obtain the approximate solution $\alpha = 0.44285$.

The boundary $x \geq 0$ gives the left-hand limit of integration. The area is then

$$\int_0^\alpha (2 - x - e^x) \cdot dx = 2 \cdot x - \frac{x^2}{2} - e^x \Big|_0^\alpha = 2 \cdot \alpha - \frac{\alpha^2}{2} - e^\alpha + 1 \approx 0.2305$$

∎

---

[3]You might wish to review Newton's Method, since it is going to be used several times in this chapter. See p.72.

3. Probability

A *random variable* is a variable that can take on a range of values and such that various sets of values have a probability assigned to them.[4] Random variables that have finitely many possible values are called *discrete*;[5] random variables taking values from an interval are called *continuous*. In statistics, continuous random variables are used all the time; even when discrete variables occur they are often approximated by continuous variables. Calculations on continuous random variables involve the calculus; we want to give typical examples. Our emphasis here is on the calculation of probability and related quantities via the integral; in a statistics course, the emphasis would be on the interpretation of those probabilities and quantities.

Probabilities for continuous random variables are computed using what is called a *probability density function*. The units of such a function are chance per unit of measurement.[6] For example, suppose that x is a continuous random variable that represents the thickness in inches of a piece of steel produced by a somewhat random process. Suppose that $9 \leq x \leq 11$ and $f(x) = 3 - (x/4)$ is the density function in chance per inch. (Inch, since that's x's units.) As we pointed out, this means that $f(x)$ is the derivative of chance with respect to x (inches), and so we compute chance (probability) by integration. For instance, the chance that $9 \leq x \leq 10$ is the integral

$$\int_9^{10} \left(3 - \frac{x}{4}\right) \cdot dx = 3 \cdot x - \frac{x^2}{8}\Big|_9^{10} = \frac{5}{8}$$

The answer 5/8 is a probability: there is a 5/8 chance that one of the pieces of steel has thickness between 9 and 11 inches. Notice that the limits of integration are the boundaries of the values of x.

---

[4]The meaning of *probability* is an interesting subject. We will take a computational approach to show how the calculus is involved.

[5]Example: a coin flip can come out in two possible ways: heads and tails.

[6]Units in ratio: a probability density function is a derivative.

If we integrate over the entire range of x, we expect to get $1 = 100\%$, since *all* values of x are between 9 and 11:

$$\int_9^{11} \left(3 - \frac{x}{4}\right) \cdot dx = 3 \cdot x - \frac{x^2}{8}\Big|_9^{11} = 1$$

Students are often surprised by the notion of a probability density function; they arise from *histograms*, which represent probabilities as areas – we might have time to say something about that in class. Our immediate purpose is to introduce a small number of famous types of continuous random variables. You are invited to search online to get an introduction to their uses, and if you take a course in statistics or research methods you will study them in detail. Our concern here is to use the integral to calculate probabilities.

Constant distribution   About as simple as you can get: the probability density function is constant on the interval $[a, b]$. What this means is that the random variable x is equally likely to any particular value on that interval. If the constant density is c, then we would need to have

$$1 = \int_a^b c \cdot dx = c \cdot x\Big|_a^b = c \cdot (b - a)$$

and this shows that $c = 1/(b - a)$. Thus, for the constant distribution, we can compute the density.

Problem T8.10. For a constant distribution on $[-3, 5]$, what is the probability that $x > 0$?

Solution.   The constant is $1/(5 + 3) = 1/8$. The probability is then

$$\int_0^5 \frac{1}{8} \cdot dx = \frac{x}{8}\Big|_0^5 = \frac{5}{8}$$

This is easy to interpret: the statement that $x > 0$ is the statement that $0 < x \leq 5$. Since that interval has length 5, and the entire interval $[-3, 5]$ has length 8, the probability that $x > 0$ is $5/8$. Sadly, when the density is not constant, we cannot use such straightforward reasoning. ■

The closed interval $[0, 5]$ computes the probability that $0 \leq x \leq 5$. We wanted $x > 0$, so we want to integrate over $0 < x \leq 5$. The chance that $x = 0$, however is $\int_0^0 (dx/8) = 0$, so the integral over $0 < x \leq 5$ is the same as that over $0 \leq x \leq 5$. There is a mild paradox here: in a continuous probability distribution, the probability that the random variables takes any particular value is 0. The probability becomes positive only over intervals.

Exponential distribution   We imagine standing by the side of a street and watching the cars go by. When a car passes, we start a stop watch, and stop it when the next car passes, recording the elapsed time. We regard the elapsed time as a continuous random variable x. The variable x is non-negative, and so its interval of values is $[0, \infty)$. The ∞ right endpoint just means that there is no right endpoint! The variable x can be arbitrarily large – theoretically.[7] Such a continuous random variable is often thought to follow the *exponential distribution*, whose probability density function looks like this:

$$(8.5) \qquad\qquad \lambda \cdot \exp(-\lambda \cdot x)$$

where λ is a positive parameter – we wil see its significance momentarily.

Since x ranges over the non-negative reals, we should have

$$(8.6) \qquad\qquad 1 = \int_0^\infty \lambda \cdot \exp(-\lambda \cdot x) \cdot dx$$

We can actually compute this integral without too much trouble. The upper limit ∞ simply means that we let the upper limit get larger and larger. Technically,

$$(8.7) \qquad \int_0^\infty \lambda \cdot \exp(-\lambda \cdot x) \cdot dx = \lim_{b \to \infty} \int_0^b \lambda \cdot \exp(-\lambda \cdot x) \cdot dx$$

An integral with an infinite upper limit is called an *improper integral*; they can be studied for their own sake, and there are several nuances that can be

---

[7]Remember that a mathematical model *represents* aspects of something else. Having infinity as an endpoint means that we don't want to worry about how large x can be.

observed. Our use of improper integrals will be confined to some probability calculations, and so we will take an ad-hoc approach, mostly in the interest of time. The integral in (8.7) with b as an endpoint can be calculated:

$$\int_0^b \lambda \cdot \exp(-\lambda \cdot x) \cdot dx = \frac{1}{-\lambda} \cdot \lambda \cdot \exp(-\lambda \cdot x) \Big|_0^b$$

$$= -\exp(-\lambda \cdot x) \Big|_0^b = 1 - \exp(-\lambda \cdot b)$$

Remembering that λ is positive, if we let b get larger and larger, the graph of the exponential function shows that $\exp(-\lambda \cdot b) \to 0$. Our improper integral becomes 1, and so, as expected, the formula (8.6) holds.

Problem T8.11. In the car watching experiment, suppose that time x is in seconds and $\lambda = 1/3$. What is the probability that two cars will pass less than 4 seconds apart?

Solution. We are asking for the probability that $0 \le x < 4$. As in a a previous problem, letting $x = 4$ does not change the answer, and so we integrate over the closed interval $[0, 4]$. We can do the integral by hand.

$$\int_0^4 \frac{1}{3} \cdot \exp\left(-\frac{x}{3}\right) \cdot dx = \frac{1}{-1/3} \cdot \frac{1}{3} \cdot \exp\left(-\frac{x}{3}\right) \Big|_0^4$$

$$= -\exp\left(-\frac{x}{3}\right) \Big|_0^4$$

$$= -\exp(-4/3) + \exp(0)$$

$$\approx 0.736$$

Our answer says that about 73.6% of the time, there will be less than 4 seconds between cars. ∎

Problem T8.12. In an exponential distribution with $\lambda = 1/5$, what is the probability that $x > 6$?

Solution. Let P be the probability we want. Remembering that x comes from $[0, \infty)$, the values $x > 6$ are in the interval $(6, \infty)$. It makes sense to compute

$$P = \int_6^\infty \frac{1}{5} \cdot \exp\left(-\frac{x}{5}\right) \cdot dx$$

We treat the infinite upper limit as we did in (8.7). The antiderivative can be done as before:

$$P = \frac{1}{-1/5} \cdot \frac{1}{5} \cdot \exp\left(-\frac{x}{5}\right) \Big|_6^\infty = -\exp\left(-\frac{x}{5}\right) \Big|_6^\infty$$

As we said, the infinite upper limit means that we think about what happens when x gets larger and larger (we can write $x \to \infty$). We see that $\exp(-x/5) \to 0$ as $x \to \infty$, and so

$$P = 0 + \exp(-6/5) \approx 0.3012$$

About 30.1% of the values of x are greater than 6. ■

   There is an interesting alternative way to do the previous problem. The values of x can be divided into two sets with no overlap: x in $[0, 6]$ and x in $(6, \infty)$. Since x has to come from one of these two intervals, if Q is the probability that x is in $[0, 6]$, and P the probability that x is in $(6, \infty)$, then $Q + P = 1$. In other words,

$$P = 1 - Q = 1 - \int_0^6 \frac{1}{5} \cdot \exp\left(-\frac{x}{5}\right) \cdot dx$$

You should compute that this gives the same answer for P as before. Notice that this alternative method avoids an infinite endpoint.

Standard normal distribution   The *standard normal distribution* occurs in a host of statistical calculations, all involving the so-called *bell-shaped curves*. The continuous random variable x ranges over the entire set of real numbers: $(-\infty, \infty)$. Here is its probability density function:

$$\frac{1}{\sqrt{2\pi}} \cdot \exp\left(-\frac{x^2}{2}\right)$$

Obviously, the $\sqrt{2\pi}$ is mysterious, not to say surprising. We would have

$$1 = \int_{-\infty}^{\infty} \frac{1}{\sqrt{2\pi}} \cdot \exp\left(-\frac{x^2}{2}\right) \cdot dx$$

This formula is quite difficult to prove theoretically. We will be content to work numerically and observe approximations, as is done in practice.

Problem T8.13. Suppose that the variable x satisfies the standard normal distribution. Find the probability that $-1 \leq x \leq 1$. (This is a famous probability: the chance of landing within one standard deviation of the mean.)

Solution. We don't know an antiderivative for $\exp(-x^2/2)$, and so we use a Riemann sum approximation to integrate the density function. We computed the sum using 5000 sub-intervals in $[-1, 1]$.

$$\frac{1}{\sqrt{2\pi}} \cdot \int_{-1}^{1} \exp\left(-\frac{x^2}{2}\right) \cdot dx \approx 0.6826$$

The reader can do an online search for a standard normal table to verify that this number is correct to the decimals shown. ∎

The Mean Many of the typical statistical quantities associated with probabilities are computed via integration. We don't want to get carried away into statistics proper, but we mention two important examples, both of which measure a kind of *average* for a continuous random variable.

Let x be a random variable in the interval $[a, b]$ with probability density function $f(x)$. (Remember that it doesn't really matter whether the endpoints a, b are included, or not. For instance, we can have $b = \infty$.) The *mean* of x is meant to be the *best approximation to x by a constant*. When we discuss *least squares* at the end of the course, we can go into this in more detail, but the idea is that the constant approximation is an *average*.[8] Here is the formula for

---

[8]The word *average* is informal.

the mean

(8.8)   mean of x $= \int_a^b x \cdot f(x) \cdot dx$

This formula is an example of the occurrence of an integral where there is no obvious anti-derivative. Riemann sums are involved here, starting with the simple formula for the mean of a finite list of numbers. We may have time to say something about this in class, but we will focus on using the integral formula.

The mean of the constant distribution is the average of the endpoints.

Problem T8.14. Show that $(a+b)/2$ is the mean of the constant distribution on $[a, b]$.

Solution. The probability density function is the constant $1/(b-a)$. The mean is then computed using (8.8).

$$\int_a^b x \cdot \frac{1}{b-a} \cdot dx = \frac{x^2}{2} \cdot \frac{1}{b-a} \Big|_a^b = \frac{1}{2(b-a)} \cdot (b^2 - a^2)$$

$$= \frac{1}{2(b-a)} \cdot (b-a)(b+a) = \frac{b+a}{2}$$

∎

In the exponential distribution with density function as in (8.5), it turns out that the mean is $1/\lambda$. (And that's the significance of the parameter λ!)

Problem T8.15. Let $\lambda = 1/5$ in the exponential distribution. Verify that the mean is 5.

Solution. Let μ be the mean. If x follows the exponential distribution with $\lambda = 1/5$, then $x \geq 0$. Thus, there is no upper limit to the values of x, and (8.8) looks like this:

$$\mu = \int_0^\infty x \cdot \frac{1}{5} \cdot \exp\left(-\frac{x}{5}\right) \cdot dx$$

We don't know how to find an antiderivative here. We experiment with Riemann sums, using higher and higher upper limit of integration to simulate going to infinity. We see that once $x \geq 300$ (or so), the integral does not change. Thus,

$$\mu \approx \int_0^{300} x \cdot \frac{1}{5} \cdot \exp\left(-\frac{x}{5}\right) \cdot dx$$

With 5000 sub-intervals, the Riemann sum is roughly 4.9994; with 10000 sub-intervals, the sum rounds to 4.99985. Close to 5, as expected. ∎

Let's use a little trickery to demonstrate an important fact:

Problem T8.16. The mean of the standard normal distribution is 0.

Solution. We defined the probability density function for this distribution. If μ is the mean, then

$$\mu = \int_{-\infty}^{\infty} x \cdot \frac{1}{\sqrt{2\pi}} \cdot \exp\left(-\frac{x^2}{2}\right) \cdot dx$$

We divide this up into two pieces: $(-\infty, 0]$ and $[0, \infty)$.

$$(8.9) \quad \mu = \int_{-\infty}^0 x \cdot \frac{1}{\sqrt{2\pi}} \cdot \exp\left(-\frac{x^2}{2}\right) \cdot dx + \int_0^{\infty} x \cdot \frac{1}{\sqrt{2\pi}} \cdot \exp\left(-\frac{x^2}{2}\right) \cdot dx$$

We will show that the first integral is the opposite of the second! To do this, let

$$F(x) = \int x \cdot \frac{1}{\sqrt{2\pi}} \cdot \exp\left(-\frac{x^2}{2}\right) \cdot dx$$

It will be convenient to have $F(0) = 0$, as well. If we are given a random antiderivative $F(x)$, we can add a constant to it to get this property. Now compute

$$\int_0^b x \cdot \frac{1}{\sqrt{2\pi}} \cdot \exp\left(-\frac{x^2}{2}\right) \cdot dx = F(b) - F(0) = F(b)$$

and we get

$$(8.10) \qquad \int_0^{\infty} x \cdot \frac{1}{\sqrt{2\pi}} \cdot \exp\left(-\frac{x^2}{2}\right) \cdot dx = \lim_{b\to\infty} F(b)$$

Use the Chain Rule to compute

$$\frac{d}{dx}F(-x) = F'(-x) \cdot (-1)$$

$$= (-1) \cdot (-x) \cdot \frac{1}{\sqrt{2\pi}} \cdot \exp\left(-\frac{(-x)^2}{2}\right)$$

$$= x \cdot \frac{1}{\sqrt{2\pi}} \cdot \exp\left(-\frac{x^2}{2}\right)$$

$$= F'(x)$$

The functions $F(-x)$ and $F(x)$ have the same derivative, and so they differ by a constant: $F(-x) = F(x) + C$ for some constant C. Let $x = 0$, and we see that $C = 0$. Thus, $F(-x) = F(x)$. Here is what follows:

$$\int_{-b}^{0} x \cdot \frac{1}{\sqrt{2\pi}} \cdot \exp\left(-\frac{x^2}{2}\right) \cdot dx = F(0) - F(-b) = -F(-b) = -F(b)$$

Thus,

(8.11) $$\int_{-\infty}^{0} x \cdot \frac{1}{\sqrt{2\pi}} \cdot \exp\left(-\frac{x^2}{2}\right) \cdot dx = \lim_{b \to \infty} -F(b)$$

Taking (8.10) and (8.11) to (8.9), we see that

$$\mu = \lim_{b \to \infty} F(b) + \lim_{b \to \infty} -F(b) = 0$$

∎

Here is a more straightforward problem.

Problem T8.17. We have a variable x with $0 \leq x \leq 1$ and the probability density is $3 \cdot x^2$. Find the mean.

Solution. From (8.8):

$$\int_{0}^{1} x \cdot 3 \cdot x^2 \cdot dx = \int_{0}^{1} 3 \cdot x^3 \cdot dx = \frac{3}{4}$$

∎

The Median For some random variables, the mean doesn't make sense as an average. An example: in the distribution of income, the mean is typically too large due to the influence of very high incomes that aren't balanced by very low incomes. It makes more sense to compute the value that divides the random variables by probability. The *median m* of a random variable x has the property that the probability of $x < m$ is 50%, and the probability that $x > m$ is 50%. The mean and median are not always the same; each gives its own idea of an average.

Problem T8.18. Find the median m for the random variable x with $0 \leq x \leq 1$ and probability density function $3 \cdot x^2$.

Solution. The probability that $0 \leq x \leq m$ is 50%:

$$\frac{1}{2} = \int_0^m 3 \cdot x^2 \cdot dx = m^3$$

and so $m = 1/\sqrt[3]{2} \approx 0.7937$. ∎

We computed the mean $\mu = 3/4$ for this distribution; the median was not the same. As we said, each of them can be used as an average depending on the context.

Here is a typical problem: the median is the solution to an algebraic equation.

Problem T8.19. Find the median of the random variable x on $[1, 3]$ with probability density $(3 \cdot x^2 + 2)/30$.

Solution. If m is the median, then

$$\frac{1}{2} = \int_1^m \frac{3 \cdot x^2 + 2}{30} \cdot dx = \frac{x^3 + 2 \cdot x}{30} \Big|_1^m$$

$$= \frac{m^3 + 2 \cdot m}{30} - \frac{3}{30}$$

Multiplying through by 30, we get

$$15 = m^3 + 2 \cdot m - 3 \quad \text{which is} \quad m^3 + 2 \cdot m - 18 = 0$$

We use Newton's Method, starting at 2, since that's in the range of the variable, and we get $m \approx 2.367$. ∎

4. Quantities in Economics

There are many quantitites that occur in economics that are computed as integrals. We survey a few of them.

Suppose we have an annual interest rate r on savings, constant for the foreseeable future.[9] Continuous compounding of savings is an exponential model, where an initial amount P dollars saved for t years results in $P \cdot \exp(r \cdot t)$ dollars at the end of the t years. Write Q for the amount at the end of t years, so that

$$Q = P \cdot \exp(r \cdot t)$$

Turning time around, we regard Q as the given and solve for P.

$$Q \cdot \exp(-r \cdot t) = P \cdot \exp(r \cdot t) \cdot \exp(-r \cdot t) = P \cdot \exp(0) = P$$

This value of P is called the *present value of Q dollars t years from now.*

(8.12)   present value of Q $= Q \cdot \exp(-r \cdot t)$

Problem T8.20. What is the present value of \$1 million 50 years from now. Assume an interest rate of 5%.

Solution.   We have $r = 0.05$ (per year, as usual), and $t = 50$ years. Recall that \$1 million is 10^6. The present value is given in (8.12)

$$\$10^6 \cdot \exp(-0.05 \cdot 50) \approx \$82,084$$

∎

It is straightforward to interpret present value. If we have \$82,084 now, and if we invest it at 5% compounded continuously, then we have \$1 million at the end of 50 years.

———————

[9]Such an interest rate is often called the *discount rate.*

Present Value of an Income Stream. We introduced the idea of an income stream on p.118: a rate at which income is accumulated. Let $f(t)$ dollars per year be an income stream, where t is time in years with $0 \leq t \leq T$. Suppose that an annual interest rate r on savings is available over that same interval. The *present value of the income stream* is the total present value of all the money from the stream. We have

$$\text{Present value of income stream} \; = \; \int_0^T f(t) \cdot \exp(-r \cdot t) \cdot dt$$

Explanation. This example was mentioned earlier in the book; the integral comes about from Riemann sums rather than by an anti-derivative. We divide the time interval $[0, T]$ into n subintervals, each of width T/n (years elapsed). Let $[p, q]$ be one of these subintervals. Over this time, the stream produces roughly $f(q) \cdot T/n$ dollars. That's since $f(t)$ measures dollars per year and T/n measures the years in $[p, q]$. The number $f(q) \cdot T/n$ only approximates the amount of money, since $f(t)$ may vary over the subinterval. The amount $f(q) \cdot T/n$ is earned about q years from now, and so its present value is given by (8.12)

$$f(q) \cdot \frac{T}{n} \cdot \exp(-r \cdot q)$$

This term has the form $f(q) \cdot \exp(-r \cdot q)$ times the width T/n of the subinterval. When these terms are added up, they form a Riemann sum for the function $f(t) \cdot \exp(-r \cdot t)$ on the interval $[0, T]$. As n gets larger and larger, (7.13) shows that the Riemann sum approaches the definite integral in the equation for the present value of the stream. ■

Problem T8.21. Find the present value of the income stream $10 \cdot t \cdot (20 - t)$ dollars per year over 20 years, if the annual interest rate is 5%.

Solution. If you graph $y = 10 \cdot t \cdot (20 - t)$ you will see that the rate of income increases over the first 10 years, and then it decreases over the final 10 years.

Notice that $5\% = 0.05$. Our formula above shows that

$$\text{Present value} = \int_0^{20} 10 \cdot t \cdot (20 - t) \cdot \exp(-0.05 \cdot t) \cdot dt$$

We used a Riemann sum to approximate this integral: ≈ 8291.07. ■

Here is the meaning of the answer: if we have \$8291 right now, we can generate the same income that the stream generates over the next 10 years. This assumes continuous compounding of interest and continuous income from the stream. Actual compounding and income would only approximate continuity.

Present value is often used to compare different ways to make the same money.

Problem T8.22. Show that each of these income streams produces a total income of \$10000 over 10 years: 1000 per year, and $200 \cdot t$ per year. Given an interest rate of 3%, which one is preferable in terms of having the smaller present value?

Solution.   The statements about income are easy to verify:

$$\int_0^{10} 1000 \cdot dt = 10000 \quad \text{and} \quad \int_0^{10} 200 \cdot t \cdot dt = 10000$$

Present value of the first stream:

$$\int_0^{10} 1000 \cdot \exp(-0.03 \cdot t) \cdot dt = \frac{1000}{0.03} \cdot (1 - e^{-0.3}) \approx \$8639.39$$

Second stream:

$$\int_0^{10} 200 \cdot t \cdot \exp(-0.03 \cdot t) \cdot dt \approx \$8208.07$$

(Obtained by the Riemann sum calculator.) The second stream has the smaller present value. ■

Is it clear why we prefer the *smaller* present value? Because it is the cheaper way to produce the income stream.

Consumer Surplus. (This is a standard topic in many texts; see [**10**], for example.[10]) A *demand curve* relates the unit price of some good to the *demand* – how many will be sold. It is natural to think of the demand x as a function of the unit price p, but our calculation will be simpler if we think of p as a function of x. So, we imagine $p = f(x)$. We usually suppose that $f(x)$ is decreasing and concave up,[11] so that $f'(x) < 0$ and $f''(x) > 0$.

Suppose that the current unit price of apples[12] is p_0, and demand is x_0, so that $p_0 = f(x_0)$. Some consumers are willing to pay more than p_0 for apples; the *consumer surplus* measures the total amount of money saved because some consumers are paying less than they are willing to pay. On the graph $p = f(x)$, these consumers are *to the left of* x_0, where $f(x)$ is *higher* than p_0.

To estimate consumer surplus, let n be a positive integer and divide the interval $[0, x_0]$ into n subintervals of width $\Delta x = x_0/n$. Let $[x_1, x_2]$ be one of these intervals, so that $x_2 - x_1 = \Delta x$. We think of this interval as standing for the Δx consumers who were willing to pay $f(x_2)$ for apples. Each of these consumers paid p_0 for the apples; each saved $f(x_2) - p_0$. Their surplus is thus $[f(x_2) - p_0] \cdot \Delta x$ (the amount each saved times the number of consumers). The total surplus for all consumers is the sum of all these terms, and that's a Riemann sum for the function $f(x) - p_0$ over the interval $[0, x_0]$.

Letting the number n get larger and larger, we arrive at the following.

$$\text{consumer surplus} \quad = \int_0^{x_0} \left[f(x) - p_0 \right] \cdot dx$$

---

[10]The concept of consumer surplus is a kind of lightning rod, as well. See the article by Currie, Murphy, and Schmitz, *The Concept of Economic Surplus and Its Use in Economic Analysis*, The Economic Journal, 1971, pp.741-799. Available on the internet.

[11]This is a form of *diminishing returns*, for the rate of decrease of p is decreasing (in absolute value).

[12]For a *unit of apples*, imagine a large crate.

We can separate out the p_0 term.

$$\int_0^{x_0} [f(x) - p_0] \cdot dx = \int_0^{x_0} f(x) \cdot dx - \int_0^{x_0} p_0 \cdot dx$$

Since p_0 is constant, the right most integral evaluates to $p_0 \cdot x_0$. We end up with this formula.

(8.13)   consumer surplus $= \int_0^{x_0} f(x) \cdot dx - p_0 \cdot x_0$

where $p = f(x)$ is the demand curve.

Problem T8.23. Suppose that a demand curve is $p = x/10 - 10 \cdot \ln(x+1) + 50$, in dollars, for $0 \le x \le 100$. At price \$15, find the consumer surplus.

Solution.   We have $p_0 = 15$, and we need to find x_0, where

$$p_0 = x_0/10 - 10 \cdot \ln(x_0 + 1) + 50$$

We need to solve the equation

$$15 = \frac{x_0}{10} - 10 \cdot \ln(x_0 + 1) + 50 \quad \text{which is} \quad \frac{x_0}{10} - 10 \cdot \ln(x_0 + 1) + 35 = 0$$

Starting at $x_0 \approx 1$, Newton's Method gives $x_0 \approx 58.357$.

  Now we use (8.13) to get the surplus C.

$$C \approx \int_0^{58.357} \left[\frac{x}{10} - 10 \cdot \ln(x+1) + 50 \right] \cdot dx - 15 \cdot 58.357 \approx 326.28$$

∎

Problem T8.24. The per unit tax on an item is supposed to generate revenue equal to 10% of the consumer surplus. If the demand curve is $p = 100 - \sqrt{x}$, in dollars, for $0 \le x \le 1000$, and the current price is \$75, what should the per item tax be?

Solution.   When $p_0 = 75$, we solve for $x_0 = 25^2 = 625$. The consumer surplus from (8.13):

$$\int_0^{625} (100 - \sqrt{x}) \cdot dx - 75 \cdot 625 \approx \$5208.33$$

The tax revenue is 10% of this surplus: $520.83. The demand is 625, and so the per item tax is

$$\frac{\$520.83}{625} \approx 83\cancel{c}$$

∎

Relating taxes to surplus is intended to limit how much taxes cut into spending. Good luck.

CHAPTER 9

Matrix Algebra

1. Matrix Arithmetic

This section introduces matrices and algebraic operations on them; perhaps it goes without saying that you will need to learn these operations.

A *matrix* is a table of numbers. Such a table organizes its entries into horizontal rows and vertical columns. The entries are often enclosed in parentheses or brackets

$$\begin{pmatrix} 1 & 2 & 3 \\ 4 & 5 & 6 \end{pmatrix} = \begin{bmatrix} 1 & 2 & 3 \\ 4 & 5 & 6 \end{bmatrix}$$

This matrix has 2 rows and 3 columns, and so its *size* is 2×3. The rows and columns are numbered in an obvious way: rows from top to bottom and columns from left to right. Thus, the second row of the given matrix is $\begin{pmatrix} 4 & 5 & 6 \end{pmatrix}$ and the first column is $\begin{pmatrix} 1 \\ 4 \end{pmatrix}$. The entry in row i and column j is called the i, j-*entry*. We often used square brackets to indicate entries: $A[1, 2]$ stands for the entry in row 1 and column 2 of matrix A. It is also common to use subscripts $A_{i,j}$, but we prefer the bracket notation, since it is more readable.

Two matrices are the same if they have the same size and the same corresponding entries. Thus, $A = B$ means that both A, B are $m \times n$, for some m, n, and also $A[i, j] = B[i, j]$ for all relevant i, j.

Matrices have an arithmetic that is somewhat similar to number arithmetic, although there are important differences that we will note.

1.1. Matrix Addition. Matrix addition is simply entry by entry. The two matrices added have to have the same size, so the entries of one will correspond to the entries of the other. For instance,

$$\begin{pmatrix} 1 & 2 & 3 \\ 4 & 5 & 6 \end{pmatrix} + \begin{pmatrix} -2 & 3 & 7 \\ 5 & -6 & -6 \end{pmatrix} = \begin{pmatrix} -1 & 5 & 10 \\ 9 & -1 & 0 \end{pmatrix}$$

It is very easy to see that addition of matrices has properties similar to those of number addition. Let A, B, C be matrices all of the same size.

(1) **Commutative Law** $A + B = B + A$

(2) **Associative Law** $(A + B) + C = A + (B + C)$

Problem T9.1. Show that the Commutative Law is true.

Solution. The definition of $A + B$ and $B + A$ shows that they are $m \times n$, since A, B are both $m \times n$. Thus, $A + B$ and $B + A$ have the same size. To compare $[i, j]$ entries, we use the definition of matrix addition to write

$$(A + B)[i, j] = A[i, j] + B[i, j]$$

Because real addition is commutative, we have $A[i, j] + B[i, j] = B[i, j] + A[i, j]$. And the definition of matrix addition shows that

$$B[i, j] + A[i, j] = (B + A)[i, j]$$

We conclude that $(A + B)[i, j] = (B + A)[i, j]$ for all i, j. The entries of $A + B$ and $B + A$ are same, and we see that they are the same matrix. ∎

In class, we will discuss the significantly different meaning of the plus sign in $(A + B)[i, j]$ and in $A[i, j] + B[i, j]$.

It is easy to take such things as the commutative and associative properties for granted; that is precisely the point of having them. These identities allow us to perform various matrix additions using familiar properties of numbers.

For each size $m \times n$ that a matrix can have, we define the matrix $\mathbb{O}_{m \times n}$ of that size and having all its entries 0. This is called the $m \times n$ *zero matrix.*

Because matrix addition is done entry by entry, it is obvious that adding the zero matrix to a matrix A does not change A.

Zero Matrix   If A is an $m \times n$ matrix, then

$$A + \mathbb{O}_{m \times n} = A = \mathbb{O}_{m \times n} + A$$

1.2. Scalar Multiplication. The next thing we want to do is to multiply a matrix by a number. When numbers are used this way, they are often called *scalars*. To multiply a matrix by a scalar, we simply multiply each entry in the matrix by the scalar. For example,

$$-2 \cdot \begin{bmatrix} 1 & 2 \\ -1 & -2 \\ 0 & 5 \end{bmatrix} = \begin{bmatrix} -2 & -4 \\ 2 & 4 \\ 0 & -10 \end{bmatrix}$$

As with matrix addition, the properties of this operation are fairly natural, and they are not hard to verify. Let A, B be matrices of the same size, and let α, β be scalars.

    (1) $0 \cdot A = \mathbb{O}$
    (2) $\beta \cdot (A + B) = (\beta \cdot A) + (\beta \cdot B)$
    (3) $(\alpha + \beta) \cdot A = (\alpha \cdot A) + (\beta \cdot A)$
    (4) $(\alpha \cdot \beta) \cdot A = \alpha \cdot (\beta \cdot A)$

We will demonstrate at least one of these in class. The argument is similar to the one we gave for matrix addition being commutative, in that we use the definition of the matrix operations and familiar properties of the real numbers.

It is customary to write $(-1) \cdot A$ as $-A$. Observe that each entry of $-A$ is the negative of the corresponding entry of A. Then $A + (-A) = \mathbb{O}$.

Matrix addition and scalar multiplication allow us to solve simple matrix equations in exactly the same way as we solve simple number equations. For example, suppose that X is an unknown matrix and we have

$$3 \cdot X + \begin{bmatrix} 1 & 2 \\ 3 & 4 \end{bmatrix} = \begin{bmatrix} 7 & 9 \\ -1 & 6 \end{bmatrix}$$

We can add the negative of a matrix to both sides to isolate $3 \cdot X$.

$$3 \cdot X + \begin{bmatrix} 1 & 2 \\ 3 & 4 \end{bmatrix} - \begin{bmatrix} 1 & 2 \\ 3 & 4 \end{bmatrix} = \begin{bmatrix} 7 & 9 \\ -1 & 6 \end{bmatrix} - \begin{bmatrix} 1 & 2 \\ 3 & 4 \end{bmatrix}$$

$$3 \cdot X + \begin{bmatrix} 0 & 0 \\ 0 & 0 \end{bmatrix} = \begin{bmatrix} 6 & 7 \\ -4 & 2 \end{bmatrix}$$

$$3 \cdot X = \begin{bmatrix} 6 & 7 \\ -4 & 2 \end{bmatrix}$$

In doing the previous calculation, we used the associative property of matrix addition and the defining property of the zero matrix. Now we scalar multiply both sides by $1/3$.

$$\frac{1}{3} \cdot 3 \cdot X = \frac{1}{3} \cdot \begin{bmatrix} 6 & 7 \\ -4 & 2 \end{bmatrix}$$

$$X = \begin{bmatrix} 2 & 7/3 \\ -4/3 & 2/3 \end{bmatrix}$$

The point of this example is that there is nothing profound involved. We solved for the matrix X as we would have solved a simple equation for a number X. We tend to use matrix arithmetic without being very self-conscious.

1.3. Matrix Multiplication. We need one more operation on matrices, a kind of multiplication. Its formula is not easy to motivate. For our immediate purposes, it will be better just to introduce multiplication and to get used to computing it.

To give the definition we start with the *dot product* of a row and a column. We need to have a row and a column with the same number of entries; we multiply corresponding entries and add up the products:

$$\begin{pmatrix} 1 & 0 & -2 & 3 \end{pmatrix} \cdot \begin{pmatrix} 3 \\ 4 \\ 1 \\ -5 \end{pmatrix} = (1 \cdot 3) + (0 \cdot 4) + (-2 \cdot 1) + (3 \cdot (-5))$$

$$= 3 + 0 - 2 - 15 = -14$$

Here is another example.

$$\begin{bmatrix} 4 & 3 & -2 \end{bmatrix} \cdot \begin{bmatrix} -1 \\ 2 \\ 1 \end{bmatrix} = 4 \cdot (-1) + 3 \cdot 2 + (-2) \cdot 1$$

$$= -4 + 6 - 2 = 0$$

Now we are ready to define matrix multiplication. There are several nuances in this definition – make sure you understand the details. First, the matrices A and B can be multiplied to form the product $A \cdot B$ *if A has the same number of columns as B has rows.* The best way to check this is to put the sizes of A and B side by side (keep A's size on the left). Suppose that A is 4×3 and B is 3×5:

$$4 \times 3 \quad 3 \times 5$$

We need the inside numbers to match: A has 3 columns and B has 3 rows. This tells us that we can form the product $A \cdot B$. The outside numbers give us the size of the result; the matrix $A \cdot B$ will be 4×5.

If A is 9×9 and B is 7×9, then, putting the sizes next to each other:

$$9 \times 9 \quad 7 \times 9$$

we see that $A \cdot B$ is *not defined*. Notice, however, that if we put B first we get

$$7 \times 9 \quad 9 \times 9$$

we see that $B \cdot A$ *is defined*! And $B \cdot A$ will be 7×9. In matrix multiplication the order of the matrices is crucial.

Let's introduce the calculation of $A \cdot B$ via an example. Suppose

$$A = \begin{bmatrix} 1 & 2 \\ 3 & 4 \\ 5 & 6 \end{bmatrix}, \quad B = \begin{bmatrix} -2 & 7 \\ 1 & -3 \end{bmatrix}$$

Looking at the sizes:

$$\begin{array}{cc} A & B \\ 3 \times 2 & 2 \times 2 \end{array}$$

we see that $A \cdot B$ is 3×2. Here is a template for a 3×2 matrix.

1,1	1,2
2,1	2,2
3,1	3,2

To get each entry in the product, we compute the dot product of a row of A and a column of B, the row and column indicated by the entry. Thus, to get the $1, 2$ entry of $A \cdot B$, we want the dot product of row 1 of A and column 2 of B:

$$(A \cdot B)[1, 2] = (\text{row 1 of } A) \cdot (\text{col 2 of } B)$$
$$= \begin{bmatrix} 1 & 2 \end{bmatrix} \cdot \begin{bmatrix} 7 \\ -3 \end{bmatrix}$$
$$= 1 \cdot 7 + 2 \cdot (-3) = 1$$

Similarly, the $3, 1$ entry of $A \cdot B$ is this.

$$(A \cdot B)[3, 1] = (\text{row 3 of } A) \cdot (\text{col 1 of } B)$$
$$= \begin{bmatrix} 5 & 6 \end{bmatrix} \cdot \begin{bmatrix} -2 \\ 1 \end{bmatrix}$$
$$= 5 \cdot (-2) + 6 \cdot 1 = -4$$

Here is the entire product.

$$A \cdot B = \begin{bmatrix} 0 & 1 \\ -2 & 9 \\ -4 & 17 \end{bmatrix}$$

Problem T9.2. Let

$$A = \begin{pmatrix} 1 & 2 & 3 & 0 \\ 4 & 5 & 6 & 0 \\ 7 & 8 & 9 & 0 \end{pmatrix} \quad \text{and} \quad B = \begin{pmatrix} 1 & -1 & 2 \\ 3 & 0 & 4 \end{pmatrix}$$

See if you can compute $A \cdot B$ and $B \cdot A$.

Solution. The matrix A has 4 columns, and B has 3 rows, so $A \cdot B$ is not defined. As for $B \cdot A$:

$$
\begin{array}{cc}
B & A \\
2 \times 3 & 3 \times 4
\end{array}
$$

Thus, $B \cdot A$ makes sense; it is 2×4.

We set up a template for the answer.

$$
\begin{pmatrix} 1 & -1 & 2 \\ 3 & 0 & 4 \end{pmatrix} \cdot \begin{pmatrix} 1 & 2 & 3 & 0 \\ 4 & 5 & 6 & 0 \\ 7 & 8 & 9 & 0 \end{pmatrix} = \begin{pmatrix} 1,1 & 1,2 & 1,3 & 1,4 \\ 2,1 & 2,2 & 2,3 & 2,4 \end{pmatrix}
$$

Here is the answer.

$$
\begin{pmatrix} 1 & -1 & 2 \\ 3 & 0 & 4 \end{pmatrix} \cdot \begin{pmatrix} 1 & 2 & 3 & 0 \\ 4 & 5 & 6 & 0 \\ 7 & 8 & 9 & 0 \end{pmatrix} = \begin{pmatrix} 11 & 13 & 15 & 0 \\ 31 & 38 & 45 & 0 \end{pmatrix}
$$

■

Multiplication is not commutative in general. As we have seen, we might have $A \cdot B$ defined but $B \cdot A$ not defined; then $A \cdot B \neq B \cdot A$. If A is 2×3 and B is 3×2, then $A \cdot B$ is 2×2, and $B \cdot A$ is 3×3, so that both $A \cdot B$ and $B \cdot A$ are defined, but they are not the same size, and so they cannot be equal. Even if $A \cdot B$ and $B \cdot A$ are both defined and of the same size, they are often not equal. As we mentioned before, in matrix multiplication the order of the matrices must be observed carefully.

The *identity matrices* act like the number 1 in multiplication. For each positive integer n, there is an $n \times n$ matrix I_n defined like this.

$$
I_2 = \begin{bmatrix} 1 & 0 \\ 0 & 1 \end{bmatrix}, \quad I_3 = \begin{bmatrix} 1 & 0 & 0 \\ 0 & 1 & 0 \\ 0 & 0 & 1 \end{bmatrix}, \quad I_4 = \begin{bmatrix} 1 & 0 & 0 & 0 \\ 0 & 1 & 0 & 0 \\ 0 & 0 & 1 & 0 \\ 0 & 0 & 0 & 1 \end{bmatrix}, \quad \cdots
$$

The matrix I_n has $I_n[i, i] = 1$ for each i; we say that the 1's are on the *diagonal*. The other entries of I_n are 0.

Identity Matrices   If A is $m \times n$, then

$$I_m \cdot A = A = A \cdot I_n$$

You should check that the size of $I_m \cdot A$ is $m \times n$ and that the size of $A \cdot I_n$ is $m \times n$. In class, we will discuss these identities; they follow directly from the formula for matrix multiplication.

We have emphasized that matrix multiplication is not commutative. However, it does have many properties similar to those of numbers. We list these properties to be clear and complete in our introduction to matrix algebra. As we mentioned previously, we will use these properties without calling attention to them. Throughout the list of rules, A, B, C are matrices.

(1) **Associative Law** If $A \cdot B$ and $B \cdot C$ are defined, then

$$(A \cdot B) \cdot C = A \cdot (B \cdot C)$$

(2) **Distributive Law** If $A \cdot B$ is defined, and if B, C have the same size, then

$$A \cdot (B + C) = (A \cdot B) + (A \cdot C)$$

(3) **Distributive Law** If $A \cdot C$ is defined, and if A, B have the same size, then

$$(A + B) \cdot C = (A \cdot C) + (B \cdot C)$$

(4) **Scalars** If $A \cdot B$ is defined, and if k is a constant, then

$$k \cdot (A \cdot B) = (k \cdot A) \cdot B = A \cdot (k \cdot B)$$

2. Applications of Matrix Algebra

In Section 3 of the Spreadsheet Appendix, we discuss the Excel formulas for matrix arithmetic. We will make use of those formulas as needed in the following.

2.1. Recursive Sequences. We consider a pair of recursive sequences, as in Chapter 2. Suppose we have A_n and B_n, for $n = 0, 1, 2, \ldots$, with these equations.

$$A_{n+1} = A_n - 2 \cdot B_n \qquad\qquad A_0 = 1$$
$$B_{n+1} = -A_n + 3 \cdot B_n \qquad\qquad B_0 = 2$$

We know that the recursive equations give us further values of both sequences. What we want to see is that the recursive equations are matrix equations:

$$(9.1) \qquad \begin{bmatrix} A_{n+1} \\ B_{n+1} \end{bmatrix} = \begin{bmatrix} 1 & -2 \\ -1 & 3 \end{bmatrix} \cdot \begin{bmatrix} A_n \\ B_n \end{bmatrix}$$

Compare this equation carefully with the recursion.

Equation (9.1) suggests using matrix multiplication to compute the sequences A_n, B_n. Here is a possible spreadsheet layout.

	A	B	C	D	E
1	1	-2			
2	-1	3			
3	n:	0	1	2	
4	A[n]	1	Λ_1	Γ_1	$\cdots$
5	B[n]	2	Λ_2	Γ_2	$\cdots$

The matrix in (9.1) appears in `A1:B2` of the spreadsheet. The initial values of the sequences appear in `B4:B5`. The entries marked with Λ_1, Λ_2 are selected when the following matrix multiplication formula is entered in `C4`:

$$\texttt{mmult(\$A\$1:\$B\$2,B4:B5)}$$

As a result, A_1, B_1 appear in `C4:C5`. If we `fill-right` from `C4:C5` to the right, we will get subsequent terms of the sequences. For instance, the entries marked Γ_1, Γ_2 will have the formula:

$$\texttt{mmult(\$A\$1:\$B\$2,C4:C5)}$$

The dollar signs around the reference to the 2×2 matrix in `A1:B2` will maintain that reference as the sequence terms are computed.

Problem T9.3. Use spreadsheet matrix multiplication to compute the first 10 terms of the sequences A_n, B_n, where

$$A_{n+1} = A_n + B_n \qquad\qquad A_0 = 4$$
$$B_{n+1} = 3 \cdot A_n - B_n \qquad\qquad B_0 = 1$$

Solution. The recursion is a matrix equation, like (9.1).

$$\begin{bmatrix} A_{n+1} \\ B_{n+1} \end{bmatrix} = \begin{bmatrix} 1 & 1 \\ 3 & -1 \end{bmatrix} \cdot \begin{bmatrix} A_n \\ B_n \end{bmatrix}$$

Here is a possible spreadsheet layout.

	A	B	C	D	E
1	matrix	1	1		
2		3	-1		
3	n:	0	1	2	
4	A[n]	4	Λ_1		
5	B[n]	1	Λ_1		

where the cells marked Λ_1 were selected when the the formula

$$\texttt{=mmult(\$B\$1:\$C\$2,B4:B5)}$$

was placed in cell `C4`. We can `fill-right` to produce the remaining 9 terms of the sequences. We get $A_9 = 1280$ and $B_9 = 2816$. ■

Here is another type of recursion that can be written as matrix multiplication.

Problem T9.4. Write a matrix equation for this recursion:

$$C_{n+2} = -6 \cdot C_{n+1} - 13 \cdot C_n \quad \text{for} \quad n = 0, 1, 2, \ldots$$

and $C_0 = 1$ and $C_1 = 0$.

Solution. The key idea is to pair C_n and C_{n+1}. Let $B_n = C_{n+1}$ for $n \geq 0$, and then $B_0 = C_1 = 0$. The recursive equation is now

$$B_{n+1} = -6 \cdot B_n - 13 \cdot C_n$$

Here are the two recursive equations for B_n and C_n:

$$C_{n+1} = B_n = 0 \cdot C_n + 1 \cdot B_n$$
$$B_{n+1} = -13 \cdot C_n - 6 \cdot B_n$$

Here is the corresponding matrix equation:

$$\begin{bmatrix} C_{n+1} \\ B_{n+1} \end{bmatrix} = \begin{bmatrix} 0 & 1 \\ -13 & -6 \end{bmatrix} \cdot \begin{bmatrix} C_n \\ B_n \end{bmatrix}$$

We can compute this sequence the way we computed the previous two. ■

2.2. Leslie's Growth Model. Patrick Leslie used matrices to model changes in animal populations over time.[1] We present a simplified example to illustrate the main idea.

We divide a female population into age categories, each category spanning a specific number of years. Each of the categories has two numbers associated with it. The *birth rate* is the average number of female offspring that an animal in that category will bear during the time for that category. The *survival rate* is the likelihood that an animal in that category will survive into the next category.

Here is a table for female New Zealand sheep.[2] Each time category is three years long.

age	birth rate	survival rate
0-3	0.436	0.795
4-6	1.502	0.787
7-9	1.513	0.462
10-12	1.313	0

Let A_n, B_n, C_n, D_n give the population in each age category after $3 \cdot n$ years.

---

[1]One of Leslie's papers on the subject is *The use of matrices in certain population mathematics* appeared in 1945 in the journal Biometrika, Vol. 33(3), pp183-212.

[2]This table is a simplified version of the more realistic (and larger) table in the paper *Parameters for Seasonally Breeding Populations*, by G. Caughley from the journal Ecology, vol. 48, 1967, pp. 834-839

Thus, A_2 is the number of female sheep aged 0-3 after 6 years; B_4 is the number of female sheep aged 4-6 after 12 years, and so on. The birth rate puts new animals into the A_n sequence. The survival rate puts animals in the *next* category. In light of this, here are recursive equations for the populations.

$$A_{n+1} = 0.436 \cdot A_n + 1.502 \cdot B_n + 1.513 \cdot C_n + 1.313 \cdot D_n$$

$$B_{n+1} = 0.795 \cdot A_n$$

$$C_{n+1} = 0.787 \cdot B_n$$

$$D_{n+1} = 0.462 \cdot C_n$$

In matrices:

$$(9.2) \qquad \begin{bmatrix} A_{n+1} \\ B_{n+1} \\ C_{n+1} \\ D_{n+1} \end{bmatrix} = \begin{bmatrix} 0.436 & 1.502 & 1.513 & 1.313 \\ 0.795 & 0 & 0 & 0 \\ 0 & 0.787 & 0 & 0 \\ 0 & 0 & 0.462 & 0 \end{bmatrix} \cdot \begin{bmatrix} A_n \\ B_n \\ C_n \\ D_n \end{bmatrix}$$

Here is an example problem.

Problem T9.5. In the sheep model just considered, suppose we start with 10 sheep in category 1 and none in the other categories. How long until the total population reaches 100 sheep?

Solution. Equation (9.2) can be used in a spreadsheet to produce the sequence, starting with

$$\begin{bmatrix} A_0 \\ B_0 \\ C_0 \\ D_0 \end{bmatrix} = \begin{bmatrix} 10 \\ 0 \\ 0 \\ 0 \end{bmatrix}$$

The total population is the sum of the populations for each category. Here are the total populations for certain values of n. (As usual, the model is approximate – this one predicts fractional numbers.)

n:	3	4	5	6
population:	37.3	59.9	97.7	158.3

The population hits 100 somewhere between $n = 5$ and $n = 6$; that's between 15 and 18 years from the start. ■

2.3. Markov Processes. These processes were introduced in Chapter 2. We have a finite number of *states*; at each time step we have to be in exactly one of the states. We are interested in *probability vectors* – a list showing the probability that we are in each of the possible states. For example, suppose there are three states: 1,2,3. A probability vector Q might look like this:

$$Q = \begin{bmatrix} 0.2 \\ 0.5 \\ 0.3 \end{bmatrix}$$

Since $Q[1] = 0.2$, the probability is 20% that we are in state 1. State 2: 50% and state 3: 30%. If we are in state 2 for certain, then $Q[2] = 1$ and $Q[1] = Q[3] = 0$. A probability vector has non-negative entries, one for each state, that add up to 1=100%, since we have to be in exactly one of the states.

The key feature of a Markov process is its *transition matrix*; that matrix shows the probability of going from one state, at a particular time, to the various states at the next time. Suppose, for example, we have three states 1,2,3, with the following transition matrix.

(9.3) $$A = \begin{bmatrix} 0.1 & 0.4 & 0.2 \\ 0.6 & 0.1 & 0.7 \\ 0.3 & 0.5 & 0.1 \end{bmatrix}$$

Each column of A imagines we are in the state of that column and tells us the chance that we move to each of the three states at the next time step. For instance, $A[3, 2] = 0.5$. If we are in state 2, then the chance that we will be in state 3 at the next time is 50%.

If Q_n holds the probability of being in each state at time n, then

(9.4) $$Q_{n+1} = A \cdot Q_n$$

gives the probabilities at time $n + 1$. We will discuss this equation in class; obviously it gives us yet another example of recursion.

It is a theorem that if the entries in A are all positive, then no matter what Q_0 is, the recursive sequence (9.4) converges to a probability distribution $\bar{Q}$ such that

$$(9.5) \qquad\qquad \bar{Q} = A \cdot \bar{Q}$$

The probability vector $\bar{Q}$ is an equilibrium for the recursion (9.4).

Problem T9.6. For the transition matrix A in (9.3) above, find the equilibrium probability vector.

Solution. The theorem to which we referred says that we can get the equilibrium probability vector $\bar{Q}$ by starting with an arbitrary probability vector Q_0 and running the recursion out. We started with

$$Q_0 = \begin{bmatrix} 1 \\ 0 \\ 0 \end{bmatrix} \quad \text{and obtained} \quad \bar{Q} \approx Q_{13} \approx \begin{bmatrix} 0.258 \\ 0.421 \\ 0.320 \end{bmatrix}$$

∎

It is instructive to start with other choices for Q_0; we should obtain the same $\bar{Q}$.

The equilibrium equation (9.5) can be written in a way that is discussed in the next chapter. Suppose that we have n states so that A is $n \times n$. Then

$$\bar{Q} = A \cdot \bar{Q} \quad \text{is} \quad I_n \cdot \bar{Q} = A \cdot \bar{Q}$$

and we can move $\bar{Q}$ to the left.

$$(9.6) \qquad\qquad (I_n - A) \cdot \bar{Q} = \mathbb{O}_{n \times 1}$$

Because $\bar{Q}$ is a probability vector, there are two more conditions: first, the entries in $\bar{Q}$ are non-negative; second, the entries in $\bar{Q}$ add up to 1. This is an example of a *system of linear equations*; those are discussed in the next chapter.

2.4. Leontief's Linear Economy. This macro-economic model is discussed in [**5**] and in [**10**]. In class, we will work with a specific example. Here, we describe the model abstractly.

We imagine an economy with n goods, and we want to keep track of the production and consumption of those goods. We have an $n \times n$ table A where $A[i, j]$ gives the number of units of good i needed to make one unit of good j. The matrix A is called the *technology matrix*. Let $X[j]$ hold the number of units of good j that is produced; we call X the *production schedule*. Then $A[i, j] \cdot X[j]$ is the number of units of good i needed to produce the $X[j]$ units of good j. Let $C[i]$ be the number of units of good i that is consumed after production; the matrix C is a *consumption vector*. Thus, the $X[i]$ units of good i are used in two ways: for production of various goods and to be consumed.

The production X has to cover both uses. In class, we will see that this leads to this inequality:

$$(9.7) \qquad\qquad A \cdot X + C \leq X$$

The meaning of inequality here is that it applies to each entry on both sides. For instance, the third entry on the left is less than or equal to the third entry on the right. And so for all entries. If (9.7) holds, we say that X is a *feasible production schedule*.

Later in the course we will see how to find a feasible production schedule, if such a schedule exists. At this point, we introduce a related concept. If we are given a feasible production schedule X_0, then we can form the recursive sequence

$$(9.8) \qquad\qquad X_{n+1} = A \cdot X_n + C \quad \text{for} \quad n = 0, 1, 2, \ldots$$

and it turns out that each X_n is a feasible production schedule. Furthermore, this sequence moves steadily toward a schedule Y that satisfies (9.7) as an equation.

(9.9) $A \cdot Y + C = Y$

The equation shows that production Y exactly covers usage in production and consumption. In other words, Y is a maximally efficient schedule.

Let's interpret equation (9.8). The right side $A \cdot X_n + C$ meaures the goods used in the economy: used for production and used for consumption. Setting X_{n+1} to this means that each producer produces next year exactly what was used this year. In many situations, when each person adjusts his own activity to make it more efficient, the efficiency of the group is unpredictable. For instance, if everyone comes early to a concert to get a good seat, the task of finding a good seat will be no easier than it was when everyone came on time. But in Leontief's model, individual adjustments lead to macro-efficiency. As we converge to (9.9), the economy becomes maximally efficient: everything that is produced is used.

Problem T9.7. We have a technology matrix A, consumption vector C, and feasible production schedule X_0, as follows

$$A = \begin{bmatrix} 0 & 0.15 & 0.225 & 0.075 \\ 1.5 & 0.075 & 0.0075 & 0.15 \\ 0 & 0.3 & 0 & 0.075 \\ 0.75 & 0.225 & 0.075 & 0.15 \end{bmatrix} \quad \text{and} \quad C = \begin{bmatrix} 30 \\ 10 \\ 100 \\ 50 \end{bmatrix} \quad \text{and} \quad X_0 = \begin{bmatrix} 150 \\ 300 \\ 250 \\ 300 \end{bmatrix}$$

Find the sequence X_n described above and estimate the schedule Y such that (9.9) holds.

Solution.   We have only to use the recursion (9.8). Excel shows that $X_{11} \approx X_{12}$, and

$$X_{12} = \begin{bmatrix} 140 \\ 286 \\ 207 \\ 277 \end{bmatrix}$$

This should be a good estimate of Y. ∎

We will discuss the spreadsheet layout of the previous problem in class. For now, we indicate what the recursion (9.8) looks like. Say, to be specific that the 4×4 matrix A appears in the spreadsheet as `A1:D4`, and suppose that C is `F1:F4`. Now suppose that X_0 occurs at `C6:C9` and that we want X_1 to appear at `D6:D9`. Then, selecting `D6:D9`, we enter this formula in `D6`:

$$\texttt{mmult(\$A\$1:\$D\$4,C6:C9)+\$F\$1:\$F\$4}$$

Using `fill-right` we can produce as many of the X_n as we wish.

CHAPTER 10

Linear Equations

1. Equations and Solutions

1.1. Definitions. In a *linear equation* a sum of multiples of the variables is set equal to a number. Such as,

$$3 \cdot x_1 - 2 \cdot x_2 + 5 \cdot x_3 + x_4 = 1$$

A *solution* to this equation is a set of values of the variables. For instance, let

$$x_1 = 1, \ x_2 = 4, \ x_3 = 0, \ x_4 = 6$$

and observe that these values make the linear equation true:[1]

$$3 \cdot 1 - 2 \cdot 4 + 5 \cdot 0 + 6 = 1$$

A *system of linear equations* is a finite list of linear equations, all in the same variables. Example: we have variables a, b, c and equations

(10.1)
$$\begin{aligned} a - 2b - c &= -2 \\ 2a + 3b + 3c &= 15 \\ a + b + c &= 6 \end{aligned}$$

Such a system can be written as a matrix product:[2]

(10.2)
$$\begin{bmatrix} 1 & -2 & -1 \\ 2 & 3 & 3 \\ 1 & 1 & 1 \end{bmatrix} \cdot \begin{bmatrix} a \\ b \\ c \end{bmatrix} = \begin{bmatrix} -2 \\ 15 \\ 6 \end{bmatrix}$$

---

[1]Don't worry about how this solution was found; we'll discuss that later.

[2]This form provides a partial explanation for the unexpected definition of matrix multiplication.

It is crucial that you see how the matrix product (10.2) corresponds directly to the equations in (10.1). There is one row of the matrix on the left for each equation. There is one column of the matrix on the left for each variable.

A *solution* to a system of equations is a solution to each of the equations simultaneously. For the system (10.1) we can let

$$\begin{bmatrix} a \\ b \\ c \end{bmatrix} = \begin{bmatrix} 3 \\ 2 \\ 1 \end{bmatrix}$$

and observe that (10.2) now gives us a true matrix equation:

$$\begin{bmatrix} 1 & -2 & -1 \\ 2 & 3 & 3 \\ 1 & 1 & 1 \end{bmatrix} \cdot \begin{bmatrix} 3 \\ 2 \\ 1 \end{bmatrix} = \begin{bmatrix} -2 \\ 15 \\ 6 \end{bmatrix}$$

as you should verify! The single matrix equation is really three equations: one for each entry – one for each of the three equations that the system requires to hold simultaneously.

Abstractly, a system of linear equations looks like this:

$$(10.3) \qquad\qquad A \cdot X = B$$

where A is a matrix of numbers – it is called the *coefficient matrix*. The symbol X stands for a column of unknowns (variables), as many variables as A has columns. The matrix B gives the right sides of each equation, and so B is called the *right side matrix*. A *solution* to this matrix equation is a specific matrix C, corresponding to X, such that

$$A \cdot C = B$$

The j-th entry of C tells the value of the j-th variable in X.

It is not hard to solve a system of linear equations. The main idea is to use one of the equations to solve for one of the variables; then you can substitute for that variable in the remaining equations, reducing the problem by one equation and one variable, and then continuing in the same way. We could take a rather ad-hoc approach to the solution of linear equations, but we choose to be systematic for at least two reasons. First, there is information hidden in the equation $AX = B$ that is relevant to general applications of matrices; a careful solution technique will disclose this information. Second, a systematic approach to solutions will allow us to reach a solution in an efficient way that avoids dead ends.

1.2. Elimination. As we mentioned above, we want to give an *official* algorithm for solving systems of linear equations, so that we can check our work, and so that we will be sure to get an answer. We will use the name *Elimination* for the algorithm. To explain it, we consider an example.

$$(10.4) \qquad \begin{pmatrix} -1 & 0 & 1 \\ 2 & 0 & 2 \\ 1 & 3 & 3 \end{pmatrix} \cdot \begin{pmatrix} a \\ b \\ c \end{pmatrix} = \begin{pmatrix} 1 \\ 2 \\ 3 \end{pmatrix}$$

Notice that each column of the coefficient matrix A corresponds to an unknown. Each row of the coefficient matrix corresponds to an equation. We put the coefficient matrix A with the right side matrix B to form a single matrix $[A|B]$. (The vertical line is just meant to separate the A from the B so that we do not confuse this juxtaposition with the matrix product AB.)

$$[A|B] = \begin{pmatrix} a & b & c & = \\ -1 & 0 & 1 & 1 \\ 2 & 0 & 2 & 2 \\ 1 & 3 & 3 & 3 \end{pmatrix}$$

The matrix $[A|B]$ is the *augmented matrix* of the system. The augmented matrix is just a table that contains *all* the numbers we are working with. We have labeled the columns of A with the unknowns, so we don't forget them! And we labeled the last column with an equals sign – it is the right side B.

Elimination involves three types of operations performed on the augmented matrix. These operations are the *elementary operations*.[3]

(1) Interchange two of the rows.

(2) Multiply a row by a non-zero number.

(3) Add a multiple of one row to another (leaving the first row the same).

Here is the key property of elementary operations: if we have matrices A, B, C and $A \cdot C = B$, and if we perform an elementary operation on the augmented matrix $[A|B]$ to obtain the matrix $[A'|B']$, then $A' \cdot C = B'$. In other words, if C is a solution to the equation $A \cdot X = B$, then C is a solution to the equation $A' \cdot X = B'$.

Furthermore, for each elementary operation there is an elementary operation that undoes it. If we interchange rows 2 and 5, say, then interchanging rows 2 and 5 again brings us back to where we started. If we multiply row 7 by 9, then multiplying row 7 by 1/9 brings us back where we started. And if we add -2 times row 3 to row 1, then adding +2 times row 3 to row 1 gets us back to where we started.[4] So, if we have the augmented matrix $[A'|B']$, as in the previous paragraph, then we can perform an elementary operation to get the matrix $[A|B]$ back again. By what we said in the previous paragraph: if C is a solution to $A' \cdot X = B'$, then C is a solution to $A \cdot X = B$.

The last two paragraphs prove this: if we are given the equation $A \cdot X = B$, and if we apply the same elementary operation to A and B to get the equation $A' \cdot X = B'$, then the second equation has *exactly the same solutions* as the first equation.

Elimination will apply elementary operations to transform a given system into a form in which the solutions (or lack of solutions) will be obvious. Since

---

[3]These operations are also called *row operations* and *elementary row operations*.

[4]The claim about this last elementary operation is a little harder to see; look at a couple of examples.

the operations do not change the set of solutions, the solutions of the given system will be obvious from the transformed system.

Let's go back to the example linear equation to demonstrate the Elimination algorithm. Then we will describe Elimination in general. In trying the algorithm on our example, observe the notation we use for the elementary operations. Rows are referred to by roman numerals. We use $-(1/3) \cdot \text{II}$ for multiplying row 2 by -1/3, and $-4 \cdot \text{I} + \text{II}$ for adding -4 times row 1 to row 2. Here is the augmented matrix we had:

$$[A|B] = \begin{pmatrix} a & b & c & = \\ -1 & 0 & 1 & 1 \\ 2 & 0 & 2 & 2 \\ 1 & 3 & 3 & 3 \end{pmatrix}$$

We begin with the -1 at position $1, 1$. This entry is called a *pivot*. It will be convenient for our pivots to be 1's, and so we multiply row one by -1 to change the pivot into a 1.

$$\begin{pmatrix} a & b & c & = \\ -1 & 0 & 1 & 1 \\ 2 & 0 & 2 & 2 \\ 1 & 3 & 3 & 3 \end{pmatrix} \quad -1 \cdot \text{I} \quad \begin{pmatrix} 1 & 0 & -1 & -1 \\ 2 & 0 & 2 & 2 \\ 1 & 3 & 3 & 3 \end{pmatrix}$$

(We drop the top row to avoid writing it over and over.) Now we use the pivot 1 in the $[1, 1]$ position to eliminate the occurrence of the variable a in all the other equations. To do this, we add multiples of row one to the other rows, clearing the entries below a. Watch, and notice that we do not change row one.

$$\begin{pmatrix} 1 & 0 & -1 & -1 \\ 2 & 0 & 2 & 2 \\ 1 & 3 & 3 & 3 \end{pmatrix} \quad \begin{matrix} -2 \cdot \text{I} + \text{II} \\ -1 \cdot \text{I} + \text{III} \end{matrix} \quad \begin{pmatrix} 1 & 0 & -1 & -1 \\ 0 & 0 & 4 & 4 \\ 0 & 3 & 4 & 4 \end{pmatrix}$$

Do you see why we added -2 times row I to row II? We wanted to clear the 2 at $[2, 1]$.

Next we move to column 2 (the column of the variable b). We already have a pivot in row one, and so we look below that row. We will use the 3 in position $[3, 2]$, since it's non-zero. We bring that entry to row two by switching rows two and three.

$$\begin{pmatrix} 1 & 0 & -1 & -1 \\ 0 & 0 & 4 & 4 \\ 0 & 3 & 4 & 4 \end{pmatrix} \quad \text{II} \leftrightarrow \text{III} \quad \begin{pmatrix} 1 & 0 & -1 & -1 \\ 0 & 3 & 4 & 4 \\ 0 & 0 & 4 & 4 \end{pmatrix}$$

Now we divide row two by the pivot 3 to turn it into a 1, as we did before.

$$\begin{pmatrix} 1 & 0 & -1 & -1 \\ 0 & 3 & 4 & 4 \\ 0 & 0 & 4 & 4 \end{pmatrix} \quad 1/3 \cdot \text{II} \quad \begin{pmatrix} 1 & 0 & -1 & -1 \\ 0 & 1 & 4/3 & 4/3 \\ 0 & 0 & 4 & 4 \end{pmatrix}$$

Next, we would eliminate variable b from the other equations, but it's already gone! We move to column 3 (the variable c), and row three. Our pivot is the 4 in $[3, 3]$. Divide row 3 by 4, and then clear above. Here goes.

$$\begin{pmatrix} 1 & 0 & -1 & -1 \\ 0 & 1 & 4/3 & 4/3 \\ 0 & 0 & 4 & 4 \end{pmatrix} \quad 1/4 \cdot \text{III} \quad \begin{pmatrix} 1 & 0 & -1 & -1 \\ 0 & 1 & 4/3 & 4/3 \\ 0 & 0 & 1 & 1 \end{pmatrix}$$

$$\begin{matrix} -4/3 \cdot \text{III} + \text{II} \\ 1 \cdot \text{III} + \text{I} \end{matrix} \quad \begin{pmatrix} 1 & 0 & 0 & 0 \\ 0 & 1 & 0 & 0 \\ 0 & 0 & 1 & 1 \end{pmatrix}$$

We have found pivots in all three rows, and so there are no more pivots to find. The resulting form is called *row-echelon form*. The row-echelon form still gives equations in a, b, c:

$$a = 0 \quad b = 0 \quad c = 1$$

Because these equations were obtained from the original equations (10.4) by elementary operations, the resulting equations have the same solutions as the original. In other words, $a = b = 0$ and $c = 1$ is the unique solution to (10.4)!

Now we describe Elimination in general, given the system $AX = B$.

1. EQUATIONS AND SOLUTIONS

Elimination

Apply Steps 1-4, with row $1, 2, \ldots$, in turn, as the *current row*, until the bottom row is reached or Step 1 fails.

Step 1. Find the leftmost column in the coefficient part of the augmented matrix having a non-zero entry at the current row or below. If there is no such entry, Step 1 fails. Otherwise, choose one such entry (this entry is a *pivot*).

Step 2. Switch rows, if necessary, to bring the pivot to the current row.

Step 3. Multiply the current row by the inverse of the pivot (so that the pivot now has value 1).

Step 4 Add multiples of the current row to rows above and below it so that the pivot is the only non-zero entry in its column. ■

Example Elimination Let's solve the system of equations given by the following augmented matrix, where the unknowns are listed along the top.

$$\begin{pmatrix} x_1 & x_2 & x_3 & x_4 & x_5 \\ 2 & -3 & 1 & 4 & 1 & 17 \\ -4 & 6 & -1 & -6 & 0 & -27 \\ 1 & 1 & 2 & 7 & -1 & 20 \\ -4 & 1 & -4 & -16 & 3 & -50 \end{pmatrix}$$

Step 1 might choose the 2 at the 1,1-entry as pivot. Step 2 is not needed, and Step 3 and Step 4 compute

$$\begin{pmatrix} 2 & -3 & 1 & 4 & 1 & 17 \\ -4 & 6 & -1 & -6 & 0 & -27 \\ 1 & 1 & 2 & 7 & -1 & 20 \\ -4 & 1 & -4 & -16 & 3 & -50 \end{pmatrix} \begin{matrix} (1/2) \cdot \mathrm{I} \\ 4 \cdot \mathrm{I} + \mathrm{II} \\ -1 \cdot \mathrm{I} + \mathrm{III} \\ 4 \cdot \mathrm{I} + \mathrm{IV} \end{matrix} \begin{pmatrix} 1 & -3/2 & 1/2 & 2 & 1/2 & 17/2 \\ 0 & 0 & 1 & 2 & 2 & 7 \\ 0 & 5/2 & 3/2 & 5 & -3/2 & 23/2 \\ 0 & -5 & -2 & -8 & 5 & -16 \end{pmatrix}$$

With row 2 as current row, the leftmost column at row 2 or below is column 2. We use $5/2$ as pivot; Step 2 interchanges rows 2 and 3:

$$\begin{pmatrix} 1 & -3/2 & 1/2 & 2 & 1/2 & 17/2 \\ 0 & 0 & 1 & 2 & 2 & 7 \\ 0 & 5/2 & 3/2 & 5 & -3/2 & 23/2 \\ 0 & -5 & -2 & -8 & 5 & -16 \end{pmatrix} \quad \text{II} \leftrightarrow \text{III} \quad \begin{pmatrix} 1 & -3/2 & 1/2 & 2 & 1/2 & 17/2 \\ 0 & 5/2 & 3/2 & 5 & -3/2 & 23/2 \\ 0 & 0 & 1 & 2 & 2 & 7 \\ 0 & -5 & -2 & -8 & 5 & -16 \end{pmatrix}$$

Then Step 3 and Step 4 come along.

$$\begin{pmatrix} 1 & -3/2 & 1/2 & 2 & 1/2 & 17/2 \\ 0 & 5/2 & 3/2 & 5 & -3/2 & 23/2 \\ 0 & 0 & 1 & 2 & 2 & 7 \\ 0 & -5 & -2 & -8 & 5 & -16 \end{pmatrix} \quad \begin{matrix} (2/5) \cdot \text{II} \\ (3/2) \cdot \text{II} + \text{I} \\ 5 \cdot \text{II} + \text{IV} \end{matrix} \quad \begin{pmatrix} 1 & 0 & 7/5 & 5 & -2/5 & 77/5 \\ 0 & 1 & 3/5 & 2 & -3/5 & 23/5 \\ 0 & 0 & 1 & 2 & 2 & 7 \\ 0 & 0 & 1 & 2 & 2 & 7 \end{pmatrix}$$

Now row 3 is the current row. We use the 1 at entry 3,3 as pivot. Step 2 and Step 3 are skipped. Step 4:

$$\begin{pmatrix} 1 & 0 & 7/5 & 5 & -2/5 & 77/5 \\ 0 & 1 & 3/5 & 2 & -3/5 & 23/5 \\ 0 & 0 & 1 & 2 & 2 & 7 \\ 0 & 0 & 1 & 2 & 2 & 7 \end{pmatrix} \quad \begin{matrix} -(3/5) \cdot \text{III} + \text{II} \\ -(7/5) \cdot \text{III} + \text{I} \\ -1 \cdot \text{III} + \text{IV} \end{matrix} \quad \begin{pmatrix} 1 & 0 & 0 & 11/5 & -16/5 & 28/5 \\ 0 & 1 & 0 & 4/5 & -9/5 & 2/5 \\ 0 & 0 & 1 & 2 & 2 & 7 \\ 0 & 0 & 0 & 0 & 0 & 0 \end{pmatrix}$$

Continue to row 4 and Step 1 fails. We are done.

Recall that the coefficient matrix you get at the end of Elimination is in *row-echelon form*; the *echelons* referred to are the columns with their pivots organized left to right and top to bottom. The number of pivots (pivoted variables) in Elimination is called the *rank* of the coefficient matrix.

The last equation in this system simply says $0 = 0$, which is always true and, therefore, has no effect on solutions. Let's write out the three equations that matter.

$$x_1 + (11/5)x_4 - (16/5)x_5 = 28/5$$
$$x_2 + (4/5)x_4 - (9/5)x_5 = 2/5$$
$$x_3 + 2x_4 + 2x_5 = 7$$

If x_4 and x_5 are chosen arbitrarily, then x_1 and x_2 and x_3 are uniquely determined in a solution to the system. Thus, there are infinitely many solutions, and, in choosing some particular solution, x_4 and x_5 are arbitrary. These arbitrary variables did not get pivots in Elimination, whereas the determined variables did get pivots. When a system is *consistent* (has solutions), the pivoted variables are determined by the non-pivoted variables. The non-pivoted variables are said to be *free* since their values are arbitrary. The pivoted variables are called *basic.*. The number of basic variables is the number of pivots is the rank of the coefficient matrix.

Problem T10.1. Solve the system of equations given by the following augmented matrix, in which the unknowns are listed along the top.

$$\begin{pmatrix} x_1 & x_2 & x_3 & = \\ 1 & 2 & 3 & 1 \\ 4 & 5 & 6 & 0 \\ 7 & 8 & 9 & 0 \end{pmatrix}$$

Solution. We put the relevant step from the Elimination algorithm in a square at the top of each cluster of calculations. As we will see, we find pivots at entries $[1, 1]$ and $[2, 2]$.

$$\begin{pmatrix} 1 & 2 & 3 & 1 \\ 4 & 5 & 6 & 0 \\ 7 & 8 & 9 & 0 \end{pmatrix} \quad \boxed{4} \atop \begin{matrix} -4 \cdot \text{I} + \text{II} \\ -7 \cdot \text{I} + \text{III} \end{matrix} \quad \begin{pmatrix} 1 & 2 & 3 & 1 \\ 0 & -3 & -6 & -4 \\ 0 & -6 & -12 & -7 \end{pmatrix}$$

$$\boxed{3} \atop -1/3 \cdot \text{II} \quad \begin{pmatrix} 1 & 2 & 3 & 1 \\ 0 & 1 & 2 & 4/3 \\ 0 & -6 & -12 & -7 \end{pmatrix} \quad \boxed{4} \atop \begin{matrix} -2 \cdot \text{II} + \text{I} \\ 6 \cdot \text{II} + \text{III} \end{matrix} \quad \begin{pmatrix} 1 & 0 & -1 & -5/3 \\ 0 & 1 & 2 & 4/3 \\ 0 & 0 & 0 & 1 \end{pmatrix}$$

We are done, since there are no further pivots in the coefficient matrix. As before, the solutions to the final equation are the same as the original. Row 3 says that

$$0x_1 + 0x_2 + 0x_3 = 1$$

and this is impossible. Thus, the equation in this problem has no solutions at all. We say the system is *inconsistent*. This example shows the general form of an inconsistent system at the end of Elimination. There will be a row of zeros in the coefficient matrix and a non-zero right side. ■

In the previous example, even though we did not find any solutions, we still get the rank of the coefficient matrix. That rank is 2. Why?

Summary   Given a system of linear equations $A \cdot X = B$, we form the augmented matrix $[A|B]$, where A is the coefficient part and B is the right side. We perform Elimination, noting which variable get pivots, and ending up with row-echelon form. If the row-echelon form has a row with 0's in the coefficient part and a non-zero right side, then the equations are inconsistent – they have no solutions. Otherwise, the equations are consistent, and the row-echelon form shows how to write the basic variables (that get pivots) in terms of the free variables (that did not get pivots). If the equations are consistent and there are no free variables, then there is a unique solution.

Here is one for you to do, perhaps in class.

Problem T10.2. Solve this system of equations.

$$-a + 2 \cdot c - d + e = -8$$
$$4 \cdot a + 2 \cdot b - 10 \cdot c - 10 \cdot e = 10$$
$$2 \cdot b - 2 \cdot c - 4 \cdot d - 6 \cdot e = -22$$
$$a + b - 3 \cdot c - 2 \cdot e = 3$$

1.3. Numerical Elimination. In class we will discuss the numerical solution of equations. This is, in general, a tricky business, but we will show how to get good information in a wide variety of problems, using the material from the next section.

2. Matrix Inverse

The numerical equation $3 \cdot x = 5$ has a simple solution $x = 5/3$. This equation comes about from multiplying both sides of $3 \cdot x = 5$ by the *inverse* of 3. If, in the linear equation $A \cdot X = E$, the matrix A has an inverse A^{-1}, then we might use that matrix in the same way:

$$A^{-1} \cdot A \cdot X = A^{-1} \cdot E \quad \text{so that} \quad X = A^{-1} \cdot E$$

It looks as if this solves the linear equation! The trouble is that not all matrices have an inverse, as we will see.

Because matrix multiplication is, in general, not commutative, we need to be pickier in the argument we just gave. Consider the equation $A \cdot X = E$, and suppose there is a matrix B such that $B \cdot A$ is an identity matrix. Let's not worry about sizes at this point, and just write $B \cdot A = I$. If $X = D$ is a solution to $A \cdot X = E$, so that $A \cdot D = E$, then we can multiply on the left by B, and get this:

$$D = I \cdot D = (B \cdot A) \cdot D = B \cdot (A \cdot D) = B \cdot E$$

This shows: *if D is a solution to $A \cdot X = E$, then $D = B \cdot E$.* In other words, the equation $A \cdot X = E$ can have at most one solution.

Observe that

(10.5)
$$\begin{bmatrix} 3 & -2 & 0 \\ 1 & 1 & 0 \end{bmatrix} \cdot \begin{bmatrix} 1 & 2 \\ 1 & 3 \\ -4 & 2 \end{bmatrix} = \begin{bmatrix} 1 & 0 \\ 0 & 1 \end{bmatrix}$$

Write the two matrix on the left side P, Q, so that $P \cdot Q = I_2$. We can apply the reasoning of the previous paragraph to Q: the equation $Q \cdot X = E$ has at most one solution. If you do Elimination on Q, you will see that it has rank 2, so that each unknown in $Q \cdot X = E$ gets a pivot and a solution would have to be unique. Sadly, the equation $Q \cdot X = E$ can be inconsistent; here is an example for you to check.

$$\begin{bmatrix} 1 & 2 \\ 1 & 3 \\ -4 & 2 \end{bmatrix} \cdot X = \begin{bmatrix} 1 \\ 0 \\ 0 \end{bmatrix}$$

We are saying that equation (10.5) does not find a solution to the present equation, since the present equation doesn't have any solution at all.

If $B \cdot A$ is an identity matrix, then $A \cdot X = E$ has *at most one solution* for each E, but sometimes there may be no solution at all.. Let's see how to get the *existence* of a solution: if there is a matrix C such that $A \cdot C$ is an identity matrix, then we claim that $C \cdot E$ is a solution to $A \cdot X = E$. Indeed,

$$A \cdot (C \cdot E) = (A \cdot C) \cdot E = I \cdot E = E$$

Going back to (10.5), we had the equation $P \cdot Q = I_2$, and so $P \cdot X = E$ has a solution for all E. If you do Elimination on P, you will see that it has rank 2, so that every row gets a pivot and there cannot be an inconsistency. However, because there is a free variable (in the third column), the solution will never be unique.

We can put the two ideas together: if there are matrices B, C such that $B \cdot A$ and $A \cdot C$ are identity matrices, then every equation $A \cdot X = E$ will have a unique solution. In this case, we say that A is *invertible*. The situation with B, C is a lot simpler than it looks, as we will see by embarking on an argument to establish a circle of facts that will tell us which matrices are invertible and how to find the B, C.

Suppose that A is invertible, and get matrices B, C so that $B \cdot A$ and $A \cdot C$ are identity matrices. Let A be $m \times n$. (Soon we will see that $m = n$, but, for now, we work in general.) The matrix $B \cdot A$ has n columns, and so I_n has to be the identity matrix equal to that product: $B \cdot A = I_n$. This shows that B is $n \times m$. Similarly, $A \cdot C$ has m rows, and so $A \cdot C = I_m$, and so C is $n \times m$.

Next we show that $B = C$! Indeed, using the associative law, we compute

$$B = B \cdot I_m = B \cdot (A \cdot C) = (B \cdot A) \cdot C = I_n \cdot C = C$$

This is important: if A is invertible, then the same, unique matrix multiplies to give an identity matrix on each side of A. We write A^{-1} for $B = C$; that matrix is the *inverse* of A. And it is clear that A is the inverse of A^{-1}.

Here is our circle of facts.

(a) Suppose that A is $m \times n$ and has an inverse. Then $m = n$ and A has rank n.

(b) Suppose that A is an $n \times n$ matrix of rank n. Then the row-echelon form of A is I_n

(c) Suppose that the matrix A has row-echelon form equal to I_n. Then A has an inverse.

Proof of (a):   Consider the equation $A \cdot X = \mathbb{O}_{m \times 1}$. As we showed above, the presence of an inverse shows that the equation $A \cdot X = \mathbb{O}$ has a unique solution. It follows that there can be no free variables in that equation, and so A has rank n. Since the rank is at most the number of rows of A, we see that $n \leq m$.

Now apply the same argument to A^{-1}, which is $n \times m$ and has inverse A. Then A^{-1} has rank m. It follows that $m \leq n$, and we have $m = n$. ∎

Proof of (b):   The first pivot goes in row 1, the second in row 2, and so on. Since the rank is n, each row gets a pivot. There are n columns, as well, and so each column gets a pivot. The pivots, which are 1's in row-echelon form, move left to right as we go down the rows, and it is easy to see that the pivots have to be at the $[i, i]$ entries for $i = 1, 2, \ldots, n$. This shows that the row-echelon form of A is I_n. ∎

Proof of (c):   Starting with the equation $A \cdot X = I_n$, we apply Elimination to $[A | I_n]$. Row-echelon form of A is I_n, and so Elimination gives this: $[I_n | C]$. The equation $I_n \cdot X = C$ has the same solutions as $A \cdot X = I_n$. We have $X = C$ is a solution to $I_n \cdot X = C$, and so we have $A \cdot C = I_n$.

Now we think about the matrix C. Since $A \cdot C = I_n$, we showed above that the equation $C \cdot X = \mathbb{O}_{n \times 1}$ has exactly one solution $X = \mathbb{O}_{n \times 1}$. Thus, in Elimination, the matrix C has rank n. By statement (b), its row-echelon form is I_n. By the first part of the argument for (c) (replacing A by C), there is a matrix D such that $C \cdot D = I_n$. This shows that C is invertible, with A, D as the matrices that give identity matrices. By what we showed above, it follows that $A = D$, and so $A \cdot C = I_n$ and $C \cdot A = I_n$. The matrix A is invertible! ■

Here is a convenient summary.

How to find an inverse   Given the $m \times n$ matrix A:

(1) If $m \neq n$, then A does not have an inverse.

(2) If $m = n$, do Elimination on $[A|I_n]$ with A as coefficient matrix. If A does not have rank n, then it does not have an inverse.

(3) If A has rank n, then the Elimination in (2) results in $[I_n|A^{-1}]$.

As mentioned in the Spreadsheet Appendix, if we are given an $n \times n$ matrix A, Excel can tell us (approximately) whether A has an inverse, and it can find the inverse, using its functions `mdeterm` and `minverse`. We keep reminding you that all numerical work is approximate; Excel can be wrong about whether a matrix has an inverse. Usually, if there is such an error, it is the error of finding an *inverse* for a matrix that doesn't actually have one. If the determinant `mdeterm` is very small, we should be careful.

3. Applications of Linear Equations

3.1. Leontief's model revisited. Recall Leontief's model economy, introduced on p.159.

Problem T10.3. In a Leontief model, let the technology matrix be A and consumption matrix C, where

$$A = \begin{bmatrix} 0.1 & 0.5 & 0.1 \\ 0.3 & 0 & 0.8 \\ 0.2 & 0.3 & 0.01 \end{bmatrix} \quad \text{and} \quad C = \begin{bmatrix} 15 \\ 20 \\ 30 \end{bmatrix}$$

Is there a production schedule X such that $A \cdot X + C = X$?

Solution. Recall that the equation for X says that it is efficient in the sense that all that is produced is used. We can place the X as an unknown in a linear equation:

$$A \cdot X + C = I_3 \cdot X \quad \text{so that} \quad C = (I_3 - A) \cdot X$$

Excel told us the the matrix $I_3 - A$ is invertible, and it computed $X = (I_3 - A)^{-1} \cdot C$:

$$X \approx \begin{bmatrix} 88 \\ 112 \\ 82 \end{bmatrix}$$

Since the entries of X are positive, these are realistic production numbers, and we have our solution. ■

3.2. Curve Fitting. We consider a general geometric problem: is there a curve of a certain sort passing through a given set of points?

Problem T10.4. Find coefficients a, b, c, so that $(1, 4)$, $(2, 4)$, and $(3, 10)$ all lie on the curve $y = a \cdot x^2 + b \cdot x + c$.

Solution. Plug the points into the equation to see what equations we need:

$$(1, 4): \quad 4 = a \cdot 1^2 + b \cdot 1 + c$$
$$(2, 5): \quad 5 = a \cdot 2^2 + b \cdot 2 + c$$
$$(3, 10): \quad 10 = a \cdot 3^2 + b \cdot 3 + c$$

We get the following system of linear equations.

(10.6)
$$\begin{bmatrix} a & b & c & \\ 1 & 1 & 1 & 4 \\ 4 & 2 & 1 & 5 \\ 9 & 3 & 1 & 10 \end{bmatrix}$$

Elimination (by hand) gives this row-echelon form.

(10.7)
$$\begin{bmatrix} 1 & 0 & 0 & 2 \\ 0 & 1 & 0 & -5 \\ 0 & 0 & 1 & 7 \end{bmatrix}$$

and our curve is $y = 2 \cdot x^2 - 5 \cdot x + 7$. ∎

The row-echelon form (10.7) of the coefficient matrix in the last problem has no rows of 0's, and so the equation (10.6) is consistent *no matter what the right side is* – no matter what are the y-values of the points considered. In other words, the existence of a solution seems to depend on the x-values.

The previous problem is a special case of a general fact about polynomials: if we have $n + 1$ points (x_i, y_i) with distinct x-coordinates, then there is a unique (polynomial) curve of the form

$$y = a_0 + a_1 \cdot x + \cdots + a_n \cdot x^n$$

passing through the points. In the curve, the coefficients a_j are the unknowns. In the case $n = 1$, we are saying that there is a unique curve $y = a_0 + a_1 \cdot x$ passing through two points with distinct x-coordinates – that's a unique line through two points. We prove the general fact in two worked-out problems.

Problem T10.5. Let $x_1, x_2, \ldots, x_{n+1}$ be distinct real numbers. Define the $(n + 1) \times (n + 1)$ matrix

$$V = \begin{bmatrix} 1 & x_1 & x_1^2 & \cdots & x_1^n \\ 1 & x_2 & x_2^2 & \cdots & x_2^n \\ \vdots & \vdots & \vdots & \ddots & \vdots \\ 1 & x_{n+1} & x_{n+1}^2 & \cdots & x_{n+1}^n \end{bmatrix}$$

Show that V has an inverse.[5]

Solution. The circle of facts on p.175 shows that V will have an inverse if its rank is $n+1$. That would mean that the equation $V \cdot X = \mathbb{O}$ would have a unique solution. The matrix X is $(n+1) \times 1$, write it

$$X = \begin{bmatrix} a_0 \\ a_1 \\ \vdots \\ a_n \end{bmatrix}$$

and we see that the equation $V \cdot X = \mathbb{O}$ amounts to this:

$$a_0 + a_1 x_1 + a_2 x_1^2 + \cdots + a_n x_1^n = 0$$
$$a_0 + a_1 x_2 + a_2 x_2^2 + \cdots + a_n x_2^n = 0$$
$$\vdots$$
$$a_0 + a_1 x_{n+1} + a_2 x_{n+1}^2 + \cdots + a_n x_{n+1}^n = 0$$

This tells us that the polynomial $a_0 + a_1 x + \cdots + a_n x^n$ has $n+1$ roots: the x_j. A polynomial of degree at most n cannot have more than n roots, unless the polynomial is the zero polynomial. In other words, all the a_j have to be 0. Thus, $V \cdot X = \mathbb{O}$ has only one solution: $X = \mathbb{O}$, and so V has rank $n+1$ and is invertible. ■

Problem T10.6. Let $(x_1, y_1), \ldots, (x_{n+1}, y_{n+1})$, where the x_j are distinct. Then there is a unique curve $y = a_0 + a_1 x + \cdots + a_n x^n$ passing through these points.

---

[5]The matrix V is called a *Vandermonde matrix*.

Solution. The fact that the curves passes through the points is that these equations hold.

$$a_0 + a_1 x_1 + \cdots + a_n x_1^n = y_1$$

$$\vdots$$

$$a_0 + a_1 x_{n+1} + \cdots + a_n x_{n+1}^n = y_n$$

The coefficient matrix for this system is the matrix V of the previous problem. By that problem, V has an inverse, and this gives us the unique curve coefficients a_i that satisfy the equations. ■

Here is a problem in which there are less than $n + 1$ points for a curve of degree n.

Problem T10.7. Find a curve $y = a_0 + a_1 x + x_2 x^2 + x_3 x^3$, where $a_3 \neq 0$, through $(2.3, 6.1)$, $(3.7, 5.3)$, $(4.2, 3)$.

Solution. The equations give the following augmented matrix.

$$\begin{bmatrix} a_0 & a_1 & a_2 & a_3 & = \\ 1 & 2.3 & 5.29 & 12.167 & 6.1 \\ 1 & 3.7 & 13.69 & 50.653 & 5.3 \\ 1 & 4.2 & 17.64 & 74.088 & 3 \end{bmatrix}$$

The coefficient matrix is 3×4, and so it cannot have an inverse. But the determinant of the first three columns is not zero, and we can therefore set a_3 to be anything we want (it is free) and solve for the other variables. Multiplying on the left by the inverse of the 3×3 matrix to the left, we get the following

$$\begin{bmatrix} a_0 & a_1 & a_2 & a_3 & = \\ 1 & 0 & 0 & 35.742 & -10.629 \\ 0 & 1 & 0 & -33.71 & 12.150 \\ 0 & 0 & 1 & 10.2 & -2.120 \end{bmatrix}$$

This discloses the infintely many solutions to the equations. If we pick $a_3 \neq 0$, then that free variable determines the basic variables a_0, a_1, a_2. ■

When the parameters are exponents, we can use the logarithm to bring them down as linear factors.

Problem T10.8. Fit the curve $w = x^a \cdot y^b \cdot z^c$, where a, b, c are parameters, to these points:

x	y	z	w
2	3	3	4
2	5	5	4
3	4	7	6

To fit the curve, get a system of linear equations from the logarithm of the equation.

Solution. Compute that

$$\ln(w) = a \cdot \ln(x) + b \cdot \ln(y) + c \cdot \ln(z)$$

We used Excel to solve the resulting system of equations, for the coefficient matrix has an inverse.

$$a \approx 2, \quad b \approx 0.724541939, \quad c \approx -0.724541939$$

■

3.3. Current in Electrical Circuits. There are many applications of linear equations to electronics. We introduce one application that generalizes to transportation networks.

Mathematically, an electrical circuit is a *discrete graph*. This is a different use of the word *graph* than in *graphing a curve*. In this context, a graph consists of finite many *vertices* and *edges*. We usually think of vertices as points; in a circuit they are *junctions*. An edge connects a pair of vertices. We usually picture the edges as line segments or curve segments; in a circuit they are the wires[6] that connect junctions. Here is a pictured example; for now, just look at the picture on the left.

---

[6]The word *wire* is euphemistic. A wire can stand for any electronic component, simple or complicated, that is part of the circuit.

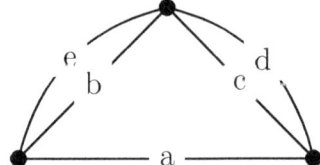   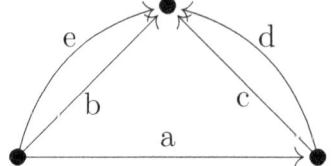

Our example has three vertices (junctions) and five edges (wires). The wires have been labeled a,b,c,d,e. Each wire has a *current* associated with it – we defined this real number on p.9; its absolute value is the number of electrons passing through that wire per unit time. To indicate the direction of the electrons, we put an arrow on each wire. That's the picture on the right. Let I_a be the current in wire a. If the current I_a is going in the direction of the arrow, then $I_a > 0$; if the current is going in the opposite direction, then $I_a < 0$. It doesn't matter which way our arrows go, the current values can be adjusted appropriately.

Now we have currents I_a, I_b, and so on. As mentioned earlier in this book, the units of current are *amperes*, usually abbreviated *amps*.

One of the basic principles of electrical circuits is called *Kirchoff's Current Rule*.[7] According to Kirchoff's Rule, the sum of the currents coming into each junction is equal to the sum of the currents leaving that junction.[8] Here are the equations for Kirchoff's Rule. They are linear equations with right side 0.

$$\text{top vertex} \quad I_e + I_b + I_c + I_d = 0$$
$$\text{left vertex:} \quad 0 = I_e + I_b + I_a$$
$$\text{right vertex:} \quad I_a = I_c + I_d$$

It turns out that we can predict which variable currents can be free in Kirchoff's equations and which can be basic. To do this, we choose a *tree* in the graph – a set of wires that does not contain a *loop* – a path from one

[7]See [**9**, p.785].

[8]Whether a current is *coming in* or *leaving* is determined by the arrow.

junction back to that same junction. (The path *does not* have to follow the arrow directions.) We give our tree as many edges as possible – it is a *maximal tree*. For instance, a, c form a maximal tree, for if we try to add an additional edge, say b, then we have a loop (b to a to c). It turns out that the variables associated with a maximal tree can be basic in Kirchoff's equations. We chose a, b, and so I_a, I_c can be basic variables, and I_b, I_d, I_e would be free. Here is the augmented matrix for the equations, putting the columns for I_a and I_c first, so they will be found basic in Elimination. The second matrix is the row-echelon form from Elimination.

$$\left[\begin{array}{ccccc|c} I_a & I_c & I_b & I_d & I_e & = \\ \hline 0 & 1 & 1 & 1 & 1 & 0 \\ -1 & 0 & -1 & 0 & -1 & 0 \\ 1 & -1 & 0 & -1 & 0 & 0 \end{array}\right] \implies \left[\begin{array}{cccccc} 1 & 0 & 1 & 0 & 1 & 0 \\ 0 & 1 & 1 & 1 & 1 & 0 \\ 0 & 0 & 0 & 0 & 0 & 0 \end{array}\right]$$

Sure enough, I_a, I_c are basic. Remember that different sets of variables can be basic. A different maximal tree would give a different set of basic variables.

3.4. Discrete Heat Distribution. We imagine a square metal plate and suppose that the temperature is held fixed along each of the four sides – a different temperature along each side. Thermal physics tells that that at equilibrium the temperature at a given point on the plate is a kind of average of the temperatures around it. We can model this using a 5 by 5 grid. We write A, B, C, D for the constant temperatures on the sides. There are 9 points interior to the grid, and we label those points by temperature variables. (Each label is below and to the left of its grid point.)

B

D

Now we interpret the condition that the temperature x_1 is the average of temperatures around it. We use the temperatures x_2, x_4, and the side temperatures C, D:

$$x_1 = \frac{1}{4} \cdot \left[x_2 + x_4 + C + D \right]$$

This gives a linear equation:

$$4x_1 - x_2 - x_4 = C + D$$

We have used $C + D$ as the right side, since C, D are given constants. Each of the nine grid points has such an equation, and we arrive at a system of nine equations in nine variables. Here is the augmented matrix of this system. Each row i comes from the average equation for grid point x_i.

$$
\begin{bmatrix}
x_1 & x_2 & x_3 & x_4 & x_5 & x_6 & x_7 & x_8 & x_9 & = \\
4 & -1 & 0 & -1 & 0 & 0 & 0 & 0 & 0 & C+D \\
-1 & 4 & -1 & 0 & -1 & 0 & 0 & 0 & 0 & D \\
0 & -1 & 4 & 0 & 0 & -1 & 0 & 0 & 0 & A+D \\
-1 & 0 & 0 & 4 & -1 & 0 & -1 & 0 & 0 & C \\
0 & -1 & 0 & -1 & 4 & -1 & 0 & -1 & 0 & 0 \\
0 & 0 & -1 & 0 & -1 & 4 & 0 & 0 & -1 & A \\
0 & 0 & 0 & -1 & 0 & 0 & 4 & -1 & 0 & B+C \\
0 & 0 & 0 & 0 & -1 & 0 & -1 & 4 & -1 & B \\
0 & 0 & 0 & 0 & 0 & -1 & 0 & -1 & 4 & A+B
\end{bmatrix}
$$

The elimination macro can be applied to the coefficient matrix to see that it has rank 9. In other words, our equations have a unique solution once A, B, C, D are given.

Our model can be any sort of graph. Here is an example we like.

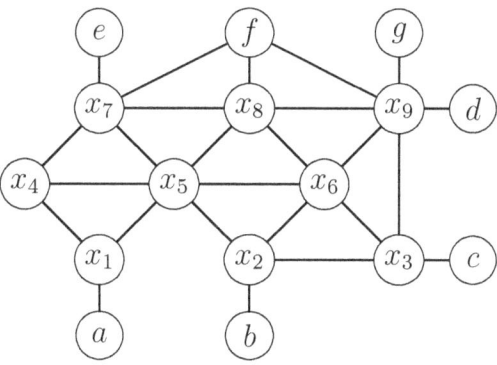

FIGURE 1. Heat Graph

The seven lettered vertices a, b, c, d, e, f, g are given values, and the variables x_j should be determined by averaging. For instance, x_1 is connected to x_4, x_5, a, and so its equation is this:

$$x_1 = \frac{1}{3} \cdot \left[x_4 + x_5 + a \right] \quad \text{which is} \quad 3 \cdot x_1 - x_4 - x_5 = a$$

Notice that x_5 is connected to six vertices, and so its value is an average of six numbers. We will write down all the equations in class or on homework.

3.5. Transshipment. We have some warehouses holding pallets of building materials. We need to ship some of the pallets around to adjust the distribution of the pallets. Here is a discrete graph of the warehouses showing allowable shipping routes between them.

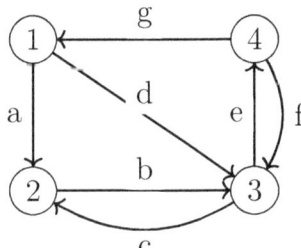

Here is a table giving the *current* supply of pallets at each vertex and the *eventual* supply, accomplished by shipping.

warehouse:	1	2	3	4
current	5	25	0	50
eventual	15	5	40	20

(Notice that there are 80 pallets total in both supply lists.) We want to determine how to ship the pallets around to get the eventual supply numbers. We have seven routes, and we let their labels stand for the amount shipped: $a, b, \ldots, g$. At warehouse 1, we have 5 pallets and we need 15, so we will have a net 10 pallets coming in. Pallets can come in along route g. Does that mean $g = 10$? Maybe we ship pallets out along routes a, d. The net number of pallets we have after shipping must be 10: $-a - d + g = 10$. We get similar equations at warehouses 2,3,4. Here are all the resulting equations.

$$-a - d + g = 10$$

$$a - b + c = -20$$

$$b - c + d - e + f = 40$$

$$e - f - g = -30$$

Here is the augmented matrix of this system.

$$\left[\begin{array}{ccccccc|c}
a & b & c & d & e & f & g & = \\
\hline
-1 & 0 & 0 & -1 & 0 & 0 & 1 & 10 \\
1 & -1 & 1 & 0 & 0 & 0 & 0 & -20 \\
0 & 1 & -1 & 1 & -1 & 1 & 0 & 40 \\
0 & 0 & 0 & 0 & 1 & -1 & -1 & -30
\end{array}\right]$$

Insight: we saw this type of matrix before when we discussed Kirchoff's equations. There, we said that we can get basic variables from a set of edges that form a maximal tree. Thus, we know we will get a solution. But because the variables represent quantities shipped, the solution needs to be non-negative. Here is the row-echelon form obtained by the Elimination macro. We omitted the last row, since it was all 0.

$$\left[\begin{array}{ccccccc|c}
a & b & c & d & e & f & g & = \\
\hline
1 & 0 & 0 & 1 & 0 & 0 & -1 & -10 \\
0 & 1 & -1 & 1 & 0 & 0 & -1 & 10 \\
0 & 0 & 0 & 0 & 1 & -1 & -1 & -30
\end{array}\right]$$

Our free variables are c, d, f, g. We want to choose non-negative values for them such that the basic variables a, b, e will also be non-negative. There are several ways to do this. For instance, we have let $c = d = f = 0$ and $g = 30$. Then $a = 20$ and $b = 40$ and $e = 0$.

There are many other applications of linear equations, some of which are introduced in the problems.

CHAPTER 11

Partial Derivatives

1. Partial derivatives

When we see an expression such as $a \cdot x^2$, we are used to thinking of x as the variable, and a as a constant. The derivative would be $2 \cdot a \cdot x$. However, since we are saying that a, x are both unknown, it makes sense to regard both of them as variables. Going back to our calculation of the derivative $2 \cdot a \cdot x$, we were thinking of x as varying and holding a constant. We might just as well let a be the variable and hold x constant. When we do that, the derivative of $a \cdot x^2$ is x^2, since $(a)' = 1$, and x^2 is a constant coefficient. To distinguish between the two possible derivatives, we call $2 \cdot a \cdot x$ the *partial derivative with respect to* x, and we call x^2 the *partial derivative with respect to* a. Here is the notation.

$$\frac{\partial}{\partial x}\left(a \cdot x^2\right) = 2 \cdot a \cdot x \quad \text{and} \quad \frac{\partial}{\partial a}\left(a \cdot x^2\right) = x^2$$

The symbol ∂ is called a *del* – it is a cursive Greek d, and it is a differential for partial derivatives.[1]

Problem T11.1. Compute

$$\frac{\partial}{\partial z}\left(y^2 + y \cdot z - e^z\right) \quad \text{and} \quad \frac{\partial}{\partial y}\left(y^2 + y \cdot z - e^z\right)$$

Solution. The partial derivative notation $\partial/\partial z$ instructs us to think of z as the variable, and to hold the other variable y constant. To help us in looking

---

[1]We feel that the distinction between the ∂ for partial derivatives and the d for ordinary derivatives is not as important as remembering what are the variables in a given context. However, most people are careful about using d only when there is one variable.

at the function this way, it might help to block out expressions in y's.

$$y^2 + y \cdot z - e^z = \bigcirc + \square \cdot z - e^z$$

Here is the derivative, with an explicit reminder of the basic rules for calculating it.

$$\frac{\partial}{\partial z}\left(\bigcirc + \square \cdot z - e^z\right) = \frac{\partial}{\partial z}\left(\bigcirc\right) + \frac{\partial}{\partial z}\left(\square \cdot z\right) - \frac{\partial}{\partial z}\left(e^z\right) \quad \text{sum rule}$$

We continue:

$$= 0 + \frac{\partial}{\partial z}\left(\square \cdot z\right) - \frac{\partial}{\partial z}\left(e^z\right) \qquad \text{derivative of constant}$$

$$= \square \cdot \frac{\partial}{\partial z}\left(z\right) - \frac{\partial}{\partial z}\left(e^z\right) \qquad \text{constant multiple rule}$$

$$= \square \cdot 1 - e^z \qquad \text{derivative of } z \text{ and } e^z$$

Now we remember that the $\square$ stood for y, and we have

$$\frac{\partial}{\partial z}\left(y^2 + y \cdot z - e^z\right) = y - e^z$$

You may or may not want to block out other variables; the point is to focus on the variable z and ignore everything else.

To compute the other derivative, we regard y as variable and hold z constant. This time, we'll leave the constant z in the expression rather than block it out. Notice that e^z will also be constant.

$$\frac{\partial}{\partial y}\left(y^2 + y \cdot z - e^z\right) = \frac{\partial}{\partial y}\left(y^2\right) + \frac{\partial}{\partial y}\left(y \cdot z\right) - \frac{\partial}{\partial y}\left(e^z\right)$$

$$= 2 \cdot y + z \cdot \frac{\partial}{\partial y}\left(y\right) - 0$$

$$= 2 \cdot y + z$$

Be sure you can explain the steps here! ■

The calculation of partial derivatives involves exactly the same rules as the calculation of ordinary derivatives; this section is essentially a review of differentiation. We say that the function f of several variables is *differentiable* if it has a valid partial derivative with respect to each of its variables.

Problem T11.2. Assume that x, t, y are independent variables. Compute

$$\frac{\partial}{\partial t} \exp(-x^3 \cdot t^2) \quad \text{and} \quad \frac{\partial}{\partial y} \exp(-x^3 \cdot t^2)$$

Solution. The partial derivative with respect to t holds x constant, and so x^3 is constant. We'll use the blocking out approach; $\square$ will stand for x^3. This derivative involves the Chain Rule:

$$\frac{\partial}{\partial t} \exp(-\square \cdot t^2) = \exp'(-\square \cdot t^2) \cdot \left(-\square \cdot t^2\right)'$$

As always, the derivative of the exponential function is itself, and so we get

$$\frac{\partial}{\partial t} \exp(-\square \cdot t^2) = \exp(-\square \cdot t^2) \cdot \left(-\square \cdot 2 \cdot t\right)$$

Now we replace the square with x^3 to get the answer.

$$\frac{\partial}{\partial t} \exp(-x^3 \cdot t^2) = \exp(-x^3 \cdot t^2) \cdot \left(-x^3 \cdot 2 \cdot t\right)$$

As for the derivative with respect to y, we hold x, t constant. Since there is no y mentioned in $\exp(-x^3 \cdot t^2)$, the entire expression is constant, and so its derivative is 0. ■

Problem T11.3. At position x on the interval $[0, 5]$, measuring cm, and at time t seconds, suppose there is heat energy $H = x \cdot (1 - x)/(1 + t)$ ergs.[2] What are the units of $\partial H/\partial x$ and of $\partial H/\partial t$? An object is at $x = 3$ when $t = 2$. In which direction should the object move to heat up? As time goes on from $t = 2$, what happens to H at $x = 3$?

Solution. The partial derivative $\partial H/\partial x$ is an ordinary derivative, with t held fixed. In other words, we have the definition of the derivative we had previously:

$$\frac{\partial H}{\partial x} = \lim_{\Delta x \to 0} \frac{\Delta H}{\Delta x} \quad \text{with} \quad \Delta t = 0$$

The units of $\Delta H/\Delta x$ are ergs per cm, and so those are the units of $\partial H/\partial x$. Similarly, the units of $\partial H/\partial t$ are ergs per second.

---

[2]The *erg* is a unit of energy, equivalent to one gram·cm^2/sec^2.

At $x = 3$ and $t = 2$ we contemplate moving along the axis with time stopped. The change in H will be measured by $\partial H/\partial x$. We write

$$H = \frac{x - x^2}{1 + t}$$

and compute

$$\frac{\partial H}{\partial x} = \frac{\partial}{\partial x}\left(\frac{x - x^2}{1 + t}\right) = \frac{1 - 2 \cdot x}{1 + t}$$

(Notice that the denominator is constant in this calculation, so we *do not use* the quotient rule. The $1 + t$ is a constant coefficient.) When $x = 3$ and $t = 2$ the partial derivative is $-5/3$ ergs per cm. Since the derivative is negative, H is *decreasing* – decreasing as x increases and t stays the same. Thus, to make H go up, we should move *to the left*.

Now we imagine keeping $x = 3$ and increasing t from 2.

$$\frac{\partial H}{\partial t} = \frac{\partial}{\partial t}\left(\frac{x - x^2}{1 + t}\right) = -\frac{x - x^2}{(1 + t)^2}$$

(For this calculation, the numerator is a constant coefficient.) When $x = 3$ and $t = 2$, the partial derivative is $6/9$ ergs per second. The heat will be increasing as time elapses from $t = 2$. ■

Problem T11.4. A plate lies over the square defined by $0 \le x \le 1$ and $0 \le y \le 1$ in the plane. (The variables x and y measure meters.) At point (x, y) on the plate, the density in kg/meter^2 is $T = x^2 - x \cdot y - 3 \cdot y^2 + 5$. Where is $\partial T/\partial x > 0$ and what does that mean?

Solution. We calculate

$$\frac{\partial T}{\partial x} = 2x - y \ \ \mathrm{kg/m}^3$$

(Make sure you understand why we get the units we do!) The inequality $\partial T/\partial x > 0$ is seen to be $2x - y > 0$, and that's $2x > y$. The inequality describes a set of points on the square – the points *below* the line $y = 2x$.

Meaning? We are thinking about changing x but not y; we are imagining a move horizontally. That the partial derivative is positive is that T increases when x increases: when we move right, the density begins to increase. ■

Let's review some rules of differentiation.

Problem T11.5. Find these derivatives.

$$\frac{\partial}{\partial z}\left(\frac{\ln|z^2 + t|}{z \cdot t + 2}\right) \quad \text{and} \quad \frac{\partial}{\partial t}\left(\frac{\ln|z^2 + t|}{z \cdot t + 2}\right)$$

Solution. We use the blocking out method on the first one, writing $\bigcirc$ for t. We need to remember that the derivative of $\ln|x|$ is $1/x$.

$$\frac{\partial}{\partial z}\left(\frac{\ln|z^2 + \bigcirc|}{z \cdot \bigcirc + 2}\right)$$

$$= \frac{(\ln|z^2 + \bigcirc|)' \cdot (z \cdot \bigcirc + 2) - \ln|z^2 + \bigcirc| \cdot (z \cdot \bigcirc + 2)'}{(z \cdot \bigcirc + 2)^2}$$

quotient rule

$$= \frac{1}{(z \cdot \bigcirc + 2)^2} \cdot \left[(\ln|z^2 + \bigcirc|)' \cdot (z \cdot \bigcirc + 2) - \ln|z^2 + \bigcirc| \cdot (z \cdot \bigcirc + 2)'\right]$$

rewrite

$$= \frac{1}{(z \cdot \bigcirc + 2)^2} \cdot \left[\frac{(z^2 + \bigcirc)'}{z^2 + \bigcirc} \cdot (z \cdot \bigcirc + 2) - \ln|z^2 + \bigcirc| \cdot (z \cdot \bigcirc + 2)'\right]$$

Chain, derivative of logarithm

$$= \frac{1}{(z \cdot \bigcirc + 2)^2} \cdot \left[\frac{2 \cdot z}{z^2 + \bigcirc} \cdot (z \cdot \bigcirc + 2) - \ln|z^2 + \bigcirc| \cdot \bigcirc\right]$$

Power Rule, constant multiple

$$= \frac{1}{(z \cdot t + 2)^2} \cdot \left[\frac{2 \cdot z}{z^2 + t} \cdot (z \cdot t + 2) - \ln|z^2 + t| \cdot t\right]$$

unblock t

We'll do the other derivative without blocking and without giving reasons. Make sure you understand each step, as always.

$$\frac{\partial}{\partial t}\left(\frac{\ln|z^2+t|}{z\cdot t+2}\right)$$

$$=\frac{(\ln|z^2+t|)'\cdot(z\cdot t+2)-\ln|z^2+t|\cdot(z\cdot t+2)'}{(z\cdot t+2)^2}$$

$$=\frac{1}{(z\cdot t+2)^2}\cdot\left[\frac{1}{z^2+t}\cdot(z\cdot t+2)-\ln|z^2+t|\cdot z\right]$$

∎

2. Higher Order Derivatives

Just as we sometimes need the second derivative of a function of a single variable, so we sometimes need to take the partial derivative more than once. This is straightforward, although the notation gets a little tricky.

For example, given $f(x,y)$, we write

$$\frac{\partial^2 f}{(\partial x)^2}=\frac{\partial}{\partial x}\cdot\frac{\partial f}{\partial x}$$

to indicate taking the partial derivative with respect to x twice.

Problem T11.6. Let $f(x,y)=(x+2\cdot y)\cdot e^{xy}$ and calculate

$$\frac{\partial^2 f}{(\partial y)^2}$$

Solution.   First we calculate

$$\frac{\partial f}{\partial y}=2\cdot e^{xy}+(x+2\cdot y)\cdot e^{xy}\cdot x$$

Then we take the partial derivative with respect to y again:

$$\frac{\partial^2 f}{(\partial y)^2}=\frac{\partial}{\partial y}\left[2\cdot e^{xy}+(x+2\cdot y)\cdot e^{xy}\cdot x\right]$$

$$=2\cdot e^{xy}\cdot x+2\cdot e^{xy}\cdot x+(x+2\cdot y)\cdot e^{xy}\cdot x^2$$

∎

Sometimes we want to change variables in the second derivative.

$$\frac{\partial^2 f}{\partial x \partial y} = \frac{\partial}{\partial x} \cdot \frac{\partial f}{\partial y}$$

Notice that the denominator differentials are applied right to left.[3]

Problem T11.7. Let $f(a, b, z) = (a + b)/(b + z^2)$. Find

$$\frac{\partial^2 f}{\partial a \partial z}$$

Solution. Compute

$$\frac{\partial}{\partial a}\left[\frac{\partial}{\partial z}\frac{a+b}{b+z^2}\right] = \frac{\partial}{\partial a}\left[-\frac{a+b}{(b+z^2)^2} \cdot 2z\right] = -\frac{1}{(b+z^2)^2} \cdot 2z$$

∎

There are times when we need more than two partial derivatives. Once you understand the notation for the second derivative, higher derivatives are not hard to interpret.

Problem T11.8. Let $G(v, x, y) = xv^3 + yv^4$. Compute

$$\frac{\partial^3 G}{\partial v \partial x \partial v}$$

Solution. We see that

$$\begin{aligned}
\frac{\partial^3 G}{\partial v \partial x \partial v} &= \frac{\partial}{\partial v} \cdot \frac{\partial}{\partial x} \cdot \frac{\partial G}{\partial v} \\
&= \frac{\partial}{\partial v} \cdot \frac{\partial}{\partial x}\left[3xv^2 + 4yv^3\right] \\
&= \frac{\partial}{\partial v}\left[3v^2\right] = 6v
\end{aligned}$$

∎

Many of the classical mathematical models are expressed via *partial differential equations*, referred to as *PDE's*. The unknown of a PDE is a function of

---

[3]It turns out that it usually doesn't matter in which order this second derivative is performed, so if you forget the ordering, you will *usually* be ok.

several variables. The PDE gives an equation that must be satisfied by various partial derivatives of the function. Here is an example.

Problem T11.9. Show that $\exp(-x \cdot y)$ is a solution z to the PDE

$$x \cdot \frac{\partial z}{\partial x} - y \cdot \frac{\partial z}{\partial y} = 0$$

Solution. The unknown z is a function of x, y; we know that by observing the partial derivatives in the equation. To show that $\exp(-x \cdot y)$ is a solution is to show that $z = \exp(-x \cdot y)$ satisfies the equation. We are being asked to plug in the function! Here goes. We will be a little breezy in calculating the partial derivatives. Make sure you can obtain the same functions.

$$\frac{\partial z}{\partial x} = \frac{\partial}{\partial x} \exp(-x \cdot y) = -y \cdot \exp(-x \cdot y)$$

and

$$\frac{\partial z}{\partial y} = \frac{\partial}{\partial y} \exp(-x \cdot y) = -x \cdot \exp(-x \cdot y)$$

Here is the left side of the PDE:

$$x \cdot \frac{\partial z}{\partial x} - y \cdot \frac{\partial z}{\partial y} = x \cdot (-y \cdot \exp(-x \cdot y)) - y \cdot (-x \cdot \exp(-x \cdot y))$$
$$= -x \cdot y \cdot \exp(-x \cdot y) + y \cdot x \cdot \exp(-x \cdot y) = 0$$

We get 0, as the equation says we should. ■

There are many famous PDE's. We won't try to learn a bunch of them, but we have chosen famous examples for the problems we will do in class and on homework.

Problem T11.10. Show that $\ln |x^2 + y^2|$ satisfies Laplace's PDE

$$\frac{\partial^2 z}{(\partial x)^2} + \frac{\partial^2 z}{(\partial y)^2} = 0$$

Solution. Remember that to *show* that something is a solution of an equation, you just plug in. We let $z = \ln|x^2 + y^2|$; the calculation goes on for a bit.

$$\frac{\partial^2}{(\partial x)^2} \ln|x^2 + y^2| = \frac{\partial}{\partial x} \frac{2 \cdot x}{x^2 + y^2}$$

$$= \frac{2 \cdot (x^2 + y^2) - 2 \cdot x \cdot (2 \cdot x)}{(x^2 + y^2)^2} \qquad = \frac{2 \cdot y^2 - 2 \cdot x^2}{(x^2 + y^2)^2}$$

The calculation of the other second derivative is similar.

$$\frac{\partial^2}{(\partial y)^2} \ln|x^2 + y^2| = \frac{2 \cdot x^2 - 2 \cdot y^2}{(x^2 + y^2)^2}$$

The two second derivatives add to 0, as needed for Laplace's PDE. ∎

3. The Chain Rule

We will need a Chain Rule for functions of several variables. There is a very general Chain Rule that actually looks very much like the one variable Chain Rule – with matrix multiplication replacing number multiplication. For our purposes, we will need only a special case of the general formula.

Suppose we have a function $f(x, y)$ and that x, y are functions of the variable t. Since t determines x, y, and x, y determine f, we see that f is a (composite) function of t. Thus, it makes sense to ask for df/dt.

Chain Rule for Several Variables If $f(x, y)$ is differentiable and if x, y are differentiable functions of t, then

(11.1)
$$\frac{df}{dt} = \frac{\partial f}{\partial x} \cdot \frac{dx}{dt} + \frac{\partial f}{\partial y} \cdot \frac{dy}{dt}$$

Explanation. To get the derivative on the left, we need to consider $\Delta f / \Delta t$. Given Δt, there are changes Δx and Δy. In class we will show how to write Δf as a sum of two changes: one with just x changing and not y, and the other with y changing but not x. We will end up with a formula

$$\Delta f = \Delta_x f + \Delta_y f$$

where the subscript indicates the variable that it changing. Then

$$\frac{\Delta f}{\Delta t} = \frac{\Delta_x f}{\Delta t} + \frac{\Delta_y f}{\Delta t}$$

Inserting Δx and Δy and ignoring that they might be 0, we write

$$\frac{\Delta f}{\Delta t} = \frac{\Delta_x f}{\Delta x} \cdot \frac{\Delta x}{\Delta t} + \frac{\Delta_y f}{\Delta y} \cdot \frac{\Delta y}{\Delta t}$$

If we let $\Delta t \to 0$, we will see (11.1) emerge.

We will give a few more details in class, but a precise argument is beyond us at this point. You should notice how the partial derivatives arise: we change f by changing x first and then y. ∎

Problem T11.11. Let $f(x, y) = x^2 + x \cdot y^3$, and $x = t^2 + 2$ and $y = 1 - 3t$. Compute df/dt.

Solution. We just imitate the Chain Rule (11.1).

$$\frac{df}{dt} = \frac{\partial}{\partial x}\left(x^2 + xy^3\right) \cdot \frac{d}{dt}\left(t^2 + 2\right) + \frac{\partial}{\partial y}\left(x^2 + xy^3\right) \cdot \frac{d}{dt}\left(1 - 3t\right)$$
$$= (2x + y^3) \cdot (2t) + 3xy^2 \cdot (-3)$$

Remembering that there are formulas for x, y in terms of t, we can write the answer as a function of t:

$$\frac{df}{dt} = \left(2(t^2 + 2) + (1 - 3t)^3\right) \cdot (2t) + 3(t^2 + 2)(1 - 3t)^2 \cdot (-3)$$

As long as we remember that x, y are functions of t, either form of the answer is fine. ∎

We have stated the Chain Rule using two variables: x, y. Any number of variables can be used, and it does not matter what they are called. Thus, if we have differentiable $f(x_1, \ldots, x_n)$, and if each x_i is a differentiable function of t, then the Chain Rule asserts that

$$\frac{df}{dt} = \frac{\partial f}{\partial x_1} \cdot \frac{dx_1}{dt} + \frac{\partial f}{\partial x_2} \cdot \frac{dx_2}{dt} + \cdots \frac{\partial f}{\partial x_n} \cdot \frac{dx_n}{dt}$$

Problem T11.12. Given $F(a, b, c)$, suppose that a, b, c are functions of the variable w. Find dF/dw.

Solution. We use a, b, c instead of x_1, x_2, x_3 and w in place of t. Then

$$\frac{dF}{dw} = \frac{\partial F}{\partial a} \cdot \frac{da}{dw} + \frac{\partial F}{\partial b} \cdot \frac{db}{dw} + \frac{\partial F}{\partial c} \cdot \frac{dc}{dw}$$

■

Here is a rather typical manipulation.

Problem T11.13. Suppose that we have $g(x, y)$ and we define $h(x) = g(x, x)$. Find $h'(x)$ in terms of the partial derivatives of g.

Solution. Let $x = t$ and $y = t$, so that x, y are functions of t. Then

$$\frac{dg}{dt} = \frac{\partial g}{\partial x} \cdot \frac{dx}{dt} + \frac{\partial g}{\partial y} \cdot \frac{dy}{dt} = \frac{\partial g}{\partial x} \cdot 1 + \frac{\partial g}{\partial y} \cdot 1$$

Since $x = t = y$, we have $g(x, y) = h(x)$ in this case. Thus, $h'(x) = dg/dt$, and so

$$h'(x) = \frac{\partial g}{\partial x} + \frac{\partial g}{\partial y}$$

■

You might try a specific instance of the previous problem. Let $g(x, y) = x + y^2$, so that $g(x, x) = x + x^2$. You can verify that

$$(x + x^2)' = \frac{\partial(x + y^2)}{\partial x} + \frac{\partial(x + y^2)}{\partial y} \quad \text{where} \quad y = x$$

Suppose we are standing at $(3, 2)$ in the xy-plane. We contemplate moving away from this point in some direction. One natural way to indicate a direction is to describe the changes in x and y that would occur in one time unit. If the changes are to be (a, b), then consider

$$x = 3 + a \cdot t$$
$$y = 2 + b \cdot t \quad \text{where} \quad t \geq 0$$

Notice that when $t = 0$ we have $x = 3, y = 2$, so we start at the point we are standing on. When $t = 1$, we have $x = 3 + a$, so that $\Delta x = a$, and we have $y = 2 + b$, so that $\Delta y = b$. Thus, a, b are the changes at time one, as we intended. As t moves from 0 to 1, we move along the line segment from $(3, 2)$ to $(3 + a, 2 + b)$.

We belabor the idea of *direction* a little more. Thinking of the x-axis and y-axis as compass directions, let's suppose we are at $(5, 6)$ and we want to go *northeast*. Here are some points that are northeast from $(5, 6)$:

$$(5.5, 6.5) \quad (6, 7) \quad (6.23, 7.23) \quad \cdots$$

It is easy to see how to get such points: add the same positive quantity to both 5 and 6. The general such point would be $(5+c, 6+c)$ for some positive number c. Notice that this idea doesn't involve time, as did the idea of the previous paragraph. That idea suggests moving along $x = 5 + c \cdot t$ and $y = 6 + c \cdot t$. The various values of c tell how far we want to go in one time unit; the larger c is, the farther we go. What this really means is that the specific c determines our *speed*.

Now suppose we have $f(x, y)$ and we want to know how f changes as we move northeast from $(5, 6)$. Let $x = 5 + c \cdot t$ and $y = 6 + c \cdot t$. Then

$$\begin{aligned}
\frac{df}{dt} &= \frac{\partial f}{\partial x} \cdot \frac{dx}{dt} + \frac{\partial f}{\partial y} \cdot \frac{dy}{dt} \\
&= \frac{\partial f}{\partial x} \cdot c + \frac{\partial f}{\partial y} \cdot c \\
&= \left(\frac{\partial f}{\partial x} + \frac{\partial f}{\partial y} \right) \cdot c
\end{aligned}$$

We see that the answer depends on c, as is not surprising. If all we are interested in is whether f increases or decreases, then since $c > 0$, we get the same sign of df/dt no matter what c we use.

Problem T11.14. Suppose that $F = x^2 \cdot y + y^3$ measures the temperature in degrees Celsius at each point (x, y) in the plane, where x, y measure meters. Suppose we are at $(1, -1)$. As we move in the direction $(-3, 4)$, what happens to the temperature?

Solution. To move from $(1, -1)$ in direction $(-3, 4)$, we let $x = 1 - 3 \cdot t$ and $y = -1 + 4 \cdot t$. The derivative dF/dt is the change in temperature with respect to t. At $t = 0$, that derivative measures the change in F at the point $(1, -1)$. Thus, we compute

$$\frac{dF}{dt} = \frac{\partial F}{\partial x} \cdot \frac{dx}{dt} + \frac{\partial F}{\partial y} \cdot \frac{dy}{dt} = 2xy \cdot (-3) + (x^2 + 3y^2) \cdot 4$$

When $t = 0$ we have $x = 1$ and $y = -1$ and $dF/dt = 22$. We see that F will increase. ∎

It is instructive to rework the previous problem using a different set of changes in the same direction as $(-3, 4)$. For instance, we can move in the direction $(-6, 8)$ or in the direction $(-1.5, 2)$. You should see that you get different values of dF/dt but that the *sign* stays the same.

Problem T11.15. The x, y-plane represents a plot of land. The equation $z = 100 - x^2 + 4x - y^2$ describes the height in yards at the point (x, y) on the land, where x, y measure yards, as well. Suppose we are standing at $(3, 2)$. In which directions (a, b) will we move down the hill?

Solution. We let

$$x = 3 + a \cdot t \quad \text{and} \quad y = 2 + b \cdot t \quad \text{for} \quad t \geq 0$$

With these equations, z is a function of x, y and x, y are functions of t, and so the Chain Rule applies:

$$\frac{dz}{dt} = (4 - 2x) \cdot a + (-2y) \cdot b$$

At the point $(3, 2)$, we see that

$$\frac{dz}{dt} = -2 \cdot a - 4 \cdot b$$

The problem asks us for directions that move down the hill: that means that $dz/dt < 0$.

$$-2 \cdot a - 4 \cdot b < 0 \quad \text{which is} \quad -2 \cdot a < 4 \cdot b \quad \text{and that's} \quad -\frac{1}{2} \cdot a < b$$

Remembering that a is change in x and b is change in y, we can graph the inequality as direction points above the line $b = -a/2$. ∎

Problem T11.16. A particle moves around the circle $x^2 + y^2 = 4$, where x, y are functions of time t. The direction of motion is defined to be

$$\begin{bmatrix} dx/dt & | & dy/dt \end{bmatrix}$$

Show that the direction of motion is always tangent to the circle.

Solution. We let $z = x^2 + y^2$, and then

$$\frac{dz}{dt} = 2x \cdot \frac{dx}{dt} + 2y \cdot \frac{dy}{dt}$$

On the other hand, $z = 4$, since we stay on the circle. Thus, $dz/dt = 0$. Therefore,

$$2x \cdot \frac{dx}{dt} + 2y \cdot \frac{dy}{dt} = 0$$

If $y \neq 0$, then we can solve for the slope of the direction:

$$\frac{dy/dt}{dx/dt} = -\frac{x}{y}$$

When (x, y) are on the circle, the slope of the radius from the center $(0, 0)$ to (x, y) is y/x. Then $-x/y$ is perpendicular to the radius: tangent to the circle.[4]

---

[4]If $x = 0$, so that y/x is not defined, then we are at one of the points $(0, \pm 2)$ on the circle, where the tangent is horizontal. In that case, $-x/y = 0$ and the direction is horizontal, as well.

If $y = 0$, then we see that $x = \pm 2$, and the Chain Rule equation says that $2x(dx/dt) = 0$. Thus, $dx/dt = 0$. The tangent to the circle at $(\pm 2, 0)$ is vertical. ∎

Given a function of several variables, its *derivative* is the matrix row of its partial derivatives. The derivative of the function f is often denoted Df; we don't usually employ the prime notation f' that was used for functions of one variable. If $f(x, y, z) = x^2 + y^2 + x \cdot z$, then the derivative Df is a 1×3 matrix:

$$Df = \begin{bmatrix} \frac{\partial f}{\partial x} & \frac{\partial f}{\partial y} & \frac{\partial f}{\partial z} \end{bmatrix} = \begin{bmatrix} 2x + z & 2y & x \end{bmatrix}$$

There isn't much to this beyond a convenient notation that we will use in the next chapter.

CHAPTER 12

Non-Linear Optimization

In Chapter 6 we considered linear optimization problems; in this chapter we consider *non-linear* problems. In general, those problems are harder: there are many more possible nuances and there is no universal algorithm for solving them, even numerically. As before, the *objective* of an optimization problem is a *function* of the variables of the problem. The typical problem also has a list of *constraints* on the variables. A constraint is an equation or inequality in one or more of the variables. A constraint is *closed* if it is an equation or a non-strict inequality (one of the forms $A = B$ and $A \leq B$ and $A \geq B$). A constraint is *open* if it is a strict inequality ($A < B$ or $A > B$).

As in Chapter 6, there are a host of applications. For example, see [**10**, Sections 8.3 and 14.2] for optimization problems in economics. In class we will discuss using the Excel `Solver` to obtain approximate solutions. Its use here is reminiscent of its use on the linear problems of the previous course, but there are important differences.

1. The First Derivative Test

We deal first with optimization problems for which all the constraints are open, or for which there are no constraints at all. Going back to the case of a single variable, the minimum or maximum of a function in an open interval occurs at a critical point – a point where the derivative is 0. Remember that the converse is false: we can have derivative zero without having an extreme. (e.g. $y = x^3$ at $x = 0$)

Throughout this chapter we will write x for a list of variables

$$x = (x_1, \ldots, x_n)$$

Then $f(x) = f(x_1, \ldots, x_n)$ will be a function of those variables. We know that $Df(x)$ is a $1 \times n$ matrix of partial derivatives. A *critical point* is a specific point p such that $Df(p) = \mathbb{O}_{1 \times n}$. This equation is really n equations, one for each of the partial derivatives:

$$\frac{\partial f}{\partial x_j}(p) = 0 \quad \text{for} \quad j = 1, 2, \ldots, n$$

Problem T12.1. Find any critical points for this function.

$$f(x_1, x_2, x_3) = x_1^3 - x_1 + 2 \cdot x_2^2 - 2 \cdot x_2 \cdot x_3 + 3 \cdot x_3^2 - 4 \cdot x_2 - 18 \cdot x_3 + 5$$

Solution. We need to solve the equation $Df = \mathbb{O}$. Here are the three equations, one for each partial derivative.

$$0 = \frac{\partial f}{\partial x_1} = 3 \cdot x_1^2 - 1$$

$$0 = \frac{\partial f}{\partial x_2} = 4 \cdot x_2 - 2 \cdot x_3 - 4$$

$$0 = \frac{\partial f}{\partial x_3} = -2 \cdot x_2 + 6 \cdot x_3 - 18$$

The first equation gives $x_1 = \pm 1/\sqrt{3}$. The second and third equations yield $x_2 = 3$, $x_3 = 4$. Thus, there are two critical points:

$$(1/\sqrt{3}, 3, 4) \quad \text{and} \quad (-1/\sqrt{3}, 3, 4)$$

As with functions of one variable, so with functions of several variables: if $f(x)$ has an extreme (maximum or minimum) at the point $x = p$, in the presence of open constraints, then $Df(p) = \mathbb{O}$. This is called the *First Derivative Test*. As with functions of one variables, we can have $Df(p) = \mathbb{O}$ without $x = p$ being an extreme.

First Derivative Test   If $f(x)$ is a differentiable function and has an extreme at $x = p$, subject to open constraints, then $Df(p) = \mathbb{O}$. ■

The fact that the logic goes only one way restricts the use of the Test.

Problem T12.2. Could $a = 1$ and $b = -2$ give the maximum of $ab^2 + 2ab$?

Solution.   We have no constraints, and so the constraints are open. Compute

$$D(ab^2 + 2ab) = \begin{bmatrix} b^2 + 2b & | & 2ab + 2a \end{bmatrix}$$

At $a = 1$ and $b = -2$, we have $D(ab^2 + 2ab) = \begin{bmatrix} 0 & -2 \end{bmatrix}$. Since this is not $\mathbb{O}$, we *cannot have a maximum*. (Or a minimum, either!) ■

The First Derivative Test suggests solving $Df(x) = \mathbb{O}$; just remember that the solutions are not necessarily meaningful. But consider the following problem.

Problem T12.3. Suppose we know that the following problem has a solution: Minimize

$$Z = 3 \cdot x^4 - 16 \cdot x^3 + 18 \cdot x^2 + 6 \cdot y^2 + 36 \cdot y + 2 \quad \text{such that} \quad x < 2$$

Find that solution.

Solution.   The constraint is open, and so the First Derivative Test applies. Compute

$$\frac{\partial Z}{\partial x} = 12 \cdot x^3 - 48 \cdot x^2 + 36 \cdot x \quad \text{and} \quad \frac{\partial Z}{\partial y} = 12 \cdot y + 36$$

To find the critical points, we set the partial derivatives equal to 0. Here is what we get: $x = 0, 1, 3$ and $y = -3$. Remembering the constraint $x < 2$, we see that $x = 0, 1$ and $y = -3$. This gives us two critical points $(0, -3)$ and $(1, -3)$. We know that the mimimum of Z has to be one of these, but which one? We have only to compare the values of Z at each point:

$$Z(0, -3) = -52, \quad Z(1, -3) = -47$$

The winner is $Z = -52$ at $(0, -3)$. ■

It was very significant that we knew there was a solution in the previous problem. Otherwise, the values we calculated at the critical points did not have to be extremes. For a problem with open constraints, this can be a tricky business. For a problem all of whose constraints are closed, there is a very important theorem that is relevant. To give that theorem, we need a term: we say that the constraints of an optimization problem are *bounded* if each variable is confined to a closed interval.

The Extreme Value Theorem   If $f(x)$ is a differentiable function subject to closed and bounded constraints, then $f(x)$ has both a maximum and minimum subject to the constraints.

Problem T12.4. Consider the problem: Maximize $x^4 \cdot y^5$, where $x^2 + 2 \cdot y^2 \leq 9$. Does the Extreme Value Theorem apply to this problem?

Solution.   The single constraint is closed; it implies that $x^2 \leq 9$, so that $-3 \leq x \leq 3$. Similarly, we have $2 \cdot y^2 \leq 9$, so that $-3/\sqrt{2} \leq y \leq 3/\sqrt{2}$, so that the constraints are bounded.[1] Thus, the Extreme Value Theorem applies. The problem necessarily has a solution. ∎

Let's take the previous problem a little further. The constraint defines an ellipse and its interior. First suppose that the maximum occurs *inside* the ellipse – where the constraint is *open*: $x^2 + 2 \cdot y^2 < 9$. Then the First Derivative Test applies and the solution occurs at a critical point, where $D(x^4 \cdot y^5) = \mathbb{O}$. On the other hand, the maximum might occur *on the ellipse* – where the constraint is $x^2 + 2 \cdot y^2 = 9$. In the next section we will discuss a method for dealing with both of these cases simultaneously.

It is instructive to use the `Solver` on the previous problem. When we started with $x = y = 0$, the `Solver` found $x = y = 0$ as solution. That's obviously not correct – perhaps the `Solver` was confused by the fact that all

---

[1]Note that we don't have to show that y can actually attain all the values between $\pm 3/\sqrt{2}$; it is enough to show that y's values have to come from that closed interval.

points on both the x-axis and y-axis are critical points. When we started $x = 1 = y$, the `Solver` found $x = 2$, $y = 1.581$ and objective 158.1 as solution. That looks more plausible. We notice that this point is on the ellipse.

We use the Exteme Value Theorem in all the problems we consider in the rest of this chapter.

2. Lagrange Multipliers

We describe *Lagrange's equation* – an equation that usually holds at the solution of an optimization problem. This equation involves the Lagrange multipliers that we introduced in Chapter 6. It is hard to motivate the equation in general, and so we will focus on using it.

2.1. Equation Constraints. Let's begin with an example.

Problem T12.5. Minimize $Z = x^2 + y^2$ where (x, y) is on the line $y = (13 - 2 \cdot x)/3$.

Solution. Notice that the single constraint is closed, but it is not bounded. Thus, the Extreme Value Theorem does not apply. We will discuss the existence of a solution later; for now, we will assume that the minimum exists.

Now to Lagrange's method. We write this constraint as a function set equal to a constant:

$$3 \cdot y + 2 \cdot x = 13$$

In this form, the constraint gets a *Lagrange multiplier* λ. (It is customary to use the Greek letter lambda as a multiplier; L for Lagrange, we think.) The multiplier is an additional variable in the problem. The Lagrange equation says, at a solution to the problem, that the derivative of the objective is the multiplier times the derivative of the function side of the constraint.

$$(12.1) \qquad DZ = \lambda \cdot D(3 \cdot y + 2 \cdot x)$$

Here are those derivatives.

$$DZ = \begin{bmatrix} 2 \cdot x & 2 \cdot y \end{bmatrix} \quad \text{and} \quad D(3 \cdot y + 2 \cdot x) = \begin{bmatrix} 2 & 3 \end{bmatrix}$$

Matrix equation (12.1) is a scalar multiple equation. Here is what it says:

$$2 \cdot x = \lambda \cdot 2$$

$$2 \cdot y = \lambda \cdot 3$$

Along with these equations, we have the line equation $3 \cdot y + 2 \cdot x = 13$. There is no one way to solve these equations. One method that *sometimes* works in harder problems is to solve for x, y in terms of λ. We get $x = \lambda$ and $y = 3 \cdot \lambda / 2$. These expressions can be substituted into the line equation:

$$3 \cdot \frac{3}{2} \cdot \lambda + 2 \cdot \lambda = 13 \quad \text{which is} \quad \lambda = 2$$

Then $x = \lambda = 2$ and $y = 3 \cdot \lambda / 2 = 3$, and we have $(x, y) = (2, 3)$, and $Z = 3^2 + 2^2 = 13$. This is the solution given by Lagrange's equation. ∎

In the previous problem, notice that Z is the square of the distance from (x, y) to $(0, 0)$, and so the point of minimum Z is the point closest to $(0, 0)$. You may remember from geometry how to find that closest point: construct a perpendicular from $(0, 0)$ to the line. The line in question has slope $-2/3$, and so a perpendicular has slope $3/2$. If we intersect the perpendicular line with the constraint line, we get the point $(2, 3)$ as the closest point. Thus, in this example we have a way to show that the Lagrange equation solution is, in fact, the actual solution to the problem. It is *usually* the case that the problem solution has to satisfy the Lagrange equation. Unfortunately, it is quite technical to give conditions for the *usually*. Most applied problems are done under the assumption that the solution satisfies the Lagrange equation, without worrying about technicalities.

In the solution to the last problem we hinted that there is no one universal method for solving the Lagrange equations. We won't do many problems by hand, in any case. Numerical solvers have their own tricks.

Let's do an example with two constraints.

Problem T12.6. Minimize $x^2 + 2 \cdot y^2 + z^2$ where $x + 3 \cdot z = y + 6$ and $2 \cdot x + y - z = -7/2$.

Solution.   Write $Q = x^2 + 2 \cdot y^2 + z^2$ for the objective. The constraints are closed; each needs to be written as a function set equal to a constraint. For the first constraint we move the y to the left. Now the function side of each constraint gets a Lagrange multiplier.

$$\lambda \quad \text{for} \quad x - y + 3 \cdot z$$
$$\mu \quad \text{for} \quad 2 \cdot x + y - z$$

Here is the Lagrange equation in this case.

$$(12.2) \qquad DQ = \lambda \cdot D(x - y + 3 \cdot z) + \mu \cdot D(2 \cdot x + y - z)$$

Taking the partial derivatives in x, y, z, in turn, to calculate the derivatives, we get these equations.

$$2 \cdot x = \lambda \cdot 1 + \mu \cdot 2 = \lambda + 2 \cdot \mu$$
$$4 \cdot y = \lambda \cdot (-1) + \mu \cdot 1 = \mu - \lambda$$
$$2 \cdot z = \lambda \cdot 3 + \mu \cdot (-1) = 3 \cdot \lambda - \mu$$

As we mentioned, there is no universal method for solving the Lagrange equations. In this problem, we solve the equations for x, y, z and then substitute into the two constraints. Here is what results.

$$x = \frac{\lambda + 2 \cdot \mu}{2} \quad y = \frac{\mu - \lambda}{4} \quad z = \frac{3 \cdot \lambda - \mu}{2}$$

$$\frac{\lambda + 2 \cdot \mu}{2} - \frac{\mu - \lambda}{4} + 3 \cdot \frac{3 \cdot \lambda - \mu}{2} = 6$$

$$2 \cdot \frac{\lambda + 2 \cdot \mu}{2} + \frac{\mu - \lambda}{4} - \frac{3 \cdot \lambda - \mu}{2} = -\frac{7}{2}$$

Clearing all the fractions we get these equations.

$$21 \cdot \lambda - 3 \cdot \mu = 24$$

$$-3 \cdot \lambda + 11 \cdot \mu = -14$$

We get $\lambda = 1$ and $\mu = -1$, which gives $x = -1/2$ and $y = -1/2$ and $z = 2$ and $Q = 19/4$. ∎

When we used the `Solver` on this problem, we obtained the same answers, approximately. Remember that the Lagrange multipliers are available on the `Sensitivity` page.

Equations (12.1) and (12.2) are special cases of the general, abstract form of Lagrange's equation. Each constraint gets a Lagrange multiplier. The derivative of the objective is the sum of the derivatives of the constraints – each constraint derivative multiplied by its multiplier.

2.2. Inequality Constraints. Inequality constraints are very common; let's show how to handle them. The short version: each constraint gets a Lagrange multiplier, as before, and Lagrange's equation holds, as before. There is an additional condition, however. Watch for it.

Problem T12.7. Maximize $x^2 + y^2$ such that $y - x^2 \leq 10$ and $x + y = 3$.

Solution. Graphing the constraints shows that they are closed and bounded, and so the Extreme Value Theorem tells us that there is a maximum. We are going to focus on the Lagrange equations, rather than on solving the problem.

We turn the inequality $y - x^2 \leq 10$ into an equation: $y - x^2 + t^2 = 10$. The variable t is a new variable being introduced temporarily; it will go away eventually. Since we have $t^2 \geq 0$, the equation $y - x^2 + t^2 = 10$ is equivalent to the inequality $y - x^2 \leq 10$. Later, we will use the fact that if $y - x^2 < 10$, then $t \neq 0$. The variable t is called a *slack variable*.

Here is the modified problem: maximize $x^2 + y^2$ such that $y - x^2 + t^2 = 10$ and $x + y = 3$. Each constraint gets a multiplier: λ_1 for $y - x^2 + t^2 = 10$ and

λ_2 for $x + y = 3$. We have three variables (x, y, t) and the Lagrange equation is this:

$$D(x^2 + y^2) = \lambda_1 \cdot D(y - x^2 + t^2) + \lambda_2 \cdot D(x + y)$$

We compute the derivatives

(12.3) $$\begin{bmatrix} 2x & 2y & 0 \end{bmatrix} = \lambda_1 \cdot \begin{bmatrix} -2x & 1 & 2t \end{bmatrix} + \lambda_2 \cdot \begin{bmatrix} 1 & 1 & 0 \end{bmatrix}$$

Look at the third entry: $0 = \lambda_1 \cdot (2t) + \lambda_2 \cdot 0 = 2 \cdot \lambda_1 \cdot t$, so that

$$0 = 2 \cdot \lambda_1 \cdot t$$

and that says that either $\lambda_1 = 0$ or $t = 0$. We noted above that if $y - x^2 < 10$, then $t \neq 0$, and so $\lambda_1 = 0$ in this case.

These conditions can be stated without reference to the slack variable t. We go back to the original variables x, y, and write (12.3) in terms of those variables alone.

(12.4) $$\begin{bmatrix} 2x & 2y \end{bmatrix} = \lambda_1 \cdot \begin{bmatrix} -2x & 1 \end{bmatrix} + \lambda_2 \cdot \begin{bmatrix} 1 & 1 \end{bmatrix}$$

And we add the condition that if $y - x^2 < 10$, then $\lambda_1 = 0$. ■

The problem we just did shows how to handle inequality constraints: they get a multiplier just as equation constraints do. If the inequality is a *proper* inequality, then the multiplier is 0.

The `Solver` found this solution to the problem:

$$x \approx 1.5, \quad y \approx 1.5, \quad \lambda_1 \approx 0, \quad \lambda_2 \approx 3$$

The maximum objective was ≈ 4.5. At the solution, the constraint $y - x^2 \leq 10$ is a proper inequality, and so its multiplier is 0. Observe that the equation (12.4) holds at the solution.

Here is a problem that involves pretty much everything we have discussed. We remind you that we will not do many problems by hand, but it will be helpful to see the Lagrange equation method in action.

Problem T12.8. Find the maximum and minimum of $2x - y$, where we have $y - x^2 \geq 0$ and $y - x \leq 2$. How do we know there is a maximum and minimum, by the way?

Solution.   Let $Q = 2x - y$, so that Q is the objective. It is not hard to see that the constraints are closed and bounded. They consist of the region between $y = x^2$ and the line $y = 2 + x$. Thus, the Extreme Value Theorem says that the maximum and minimum exist.

Let λ_1 be the multiplier for $y - x^2 \geq 0$ and let λ_2 be the multiplier for $y - x \leq 2$. Then

$$D(2x - y) = \lambda_1 \cdot D(y - x^2) + \lambda_2 \cdot D(y - x)$$
$$\begin{bmatrix} 2 & -1 \end{bmatrix} = \lambda_1 \cdot \begin{bmatrix} -2x & 1 \end{bmatrix} + \lambda_2 \cdot \begin{bmatrix} -1 & 1 \end{bmatrix}$$

We also have these conditions: if $y - x^2 > 0$, then $\lambda_1 = 0$ and if $y - x < 2$, then $\lambda_2 = 0$.

The conditions suggest cases:

Case 1   Let $y - x^2 > 0$ and $y - x < 2$.

We have $\lambda_1 = 0 = \lambda_2$, and the Lagrange equation is

$$\begin{bmatrix} 2 & -1 \end{bmatrix} = \begin{bmatrix} 0 & 0 \end{bmatrix}$$

That is impossible: there are no points given in this case.

Case 2   Let $y - x^2 = 0$ and $y - x < 2$.

We have $\lambda_2 = 0$, and the equation:

$$\begin{bmatrix} 2 & -1 \end{bmatrix} = \lambda_1 \cdot \begin{bmatrix} -2x & 1 \end{bmatrix}$$

The second entry says that $\lambda = -1$, and the first entry equation is $2 = (-1) \cdot (-2x) = 2x$. Thus, $x = 1$, and so $y - x^2 = 0$ says that $y = 1$. The objective is $Q = 1$ in this case.

Case 3 Let $y - x^2 > 0$ and $y - x = 2$.

In this case, $\lambda_1 = 0$ and the equation is

$$\begin{bmatrix} 2 & -1 \end{bmatrix} = \lambda_2 \cdot \begin{bmatrix} -1 & 1 \end{bmatrix}$$

The entries disagree on the value of λ_2. There are no points in this case.

Case 4 Let $y - x^2 = 0$ and $y - x = 2$.

We substitute $y = x^2$ into the second equation: $x^2 - x = 2$, so that $x^2 - x - 2 = 0$ and that's $(x - 2)(x + 1) = 0$. Then $x = 2, -1$. When $x = 2$, $y = 4$, and the objective is $Q = 0$. When $x = -1$, we have $y = 1$ and the objective is $Q = -3$.

Looking at all the cases, we see that the maximum objective is 1, at $(1, 1)$, and the minimum objective is -3 at $(-1, 1)$. ■

Notice, in Case 4, when $x = 2$ and $y = 4$, the Lagrange equations are satisfied, but this point gives neither a maximum nor a minimum for the objective.

The `Solver` finds both the maximum and minimum, and it finds the Lagrange multipliers in each case. It is interesting to observe that Lagrange's equation holds in each case.

When do the Lagrange equations hold?

As we mentioned, this issue is usually ignored. And it's somewhat complicated to give a complete set of conditions under which the equations have to hold. Here is a fairly simple criterion: assume we have k equation constraints:

$$g_1(x, y, z) = c_1$$
$$g_2(x, y, z) = c_2$$
$$\vdots$$
$$g_k(x, y, z) = c_k$$

where $c_1, \ldots, c_k$ are constants. We have used three variables; it does not matter how many we have.

We bundle the derivative of each g_j into one matrix:

$$G = \begin{bmatrix} Dg_1 \\ Dg_2 \\ \vdots \\ Dg_k \end{bmatrix}$$

If the rank of G is k at the maximum or minimum of the objective, then the Lagrange equations have to hold. Thus, to conduct a complete use of the Lagrange equations, we need to check points where Dg has rank less than k, plugging them into the objective to see if the minimum/maximum occurs. Typically, there are no such points, but, to be complete, this case should be considered.

2.3. Shadow Prices. Just as in the case of linear optimization problems, the Lagrange multipliers tell what happens when a parameter is changed. Let's see how this works in a simple problem.

Let p be a constant initially 16, and we want to maximize the product of two positive numbers whose sum is p.

If x, y are the two positive numbers, our objective $Z = x \cdot y$. We have $x + y = p$ and $x, y > 0$. In getting the Lagrange equations, we only deal with the closed constraint. Let λ be the multiplier, and the Lagrange equation is this:

$$DZ = \lambda \cdot D(x + y)$$

and that's

$$y = \lambda \quad \text{and} \quad x = \lambda$$

The constraint is then $\lambda + \lambda = p$, so that

(12.5) $$\lambda = \frac{p}{2}$$

This gives $x = p/2$ and $y = p/2$ and $Z = p^2/4$. Notice that x, y are positive, so we have satisfied all the constraints.

We write $\bar{Z} = p^2/4$ for the maximum value of Z. The shadow price of the constraint $x + y = p$ is the derivative

$$\frac{d\bar{Z}}{dp} = \frac{p}{2}$$

We said that $p = 16$ initially; that means that we expect $\bar{Z}$ to increase by 8 units for each unit increase in p.

Equation (12.5) shows that the shadow price is the Lagrange multiplier. We claim that this is what happens in general. In linear optimization problems, the shadow price is the *exact* amount of increase in the optimum objective. In non-linear problems, it is an approximation.

Summary: In an optimization problem, we imagine that the constant side of one of the constraints is a parameter that might change. Then the Lagrange multiplier for that constraint, at a solution to the problem, gives the derivative of the optimal objective with respect to that parameter.

Problem T12.9. We sell two related products A and B. Here are the demand curves for each; they show the demand as a function of the price. The subscript 1 goes with A, 2 goes with B.

$$x_1 = 6 \cdot 10^5 \cdot \frac{p_2}{p_2 + 400} \cdot \frac{1}{p_1^2 + 10} \quad \text{and} \quad x_2 = 50 \cdot \exp(-p_2/200)$$

It takes 10 hours to make one unit of A and 40 hours for one unit of B. We have up to 1000 hours for production of A and 500 hours for B.

(a) Find the prices and demand that maximize revenue.

(b) What would we be willing to pay for 10 more hours for producing A?

Solution. The objective is revenue R, where

$$R = x_1 \cdot p_1 + x_2 \cdot p_2$$

Constraints? The two demand functions, and the production inequalities:

$$10 \cdot x_1 \le 1000 \quad \text{and} \quad 40 \cdot x_2 \le 500$$

In the `Solver` we set the variables p_1, p_2 to be non-negative. Our solution was this.

$$p_1 \approx 49.88 \quad x_1 \approx 100 \quad p_2 \approx 285.33 \quad x_2 \approx 12$$

with maximum revenue $\bar{R} \approx 8414$.

Regarding as a parameter h the 1000 hours for production of A, an additional 10 hours is $\Delta h = 10$. The `Solver` tells us that the Lagrange multiplier of the constraint $10 \cdot x_1 \le h$ is 2.484. Thus

$$\frac{\Delta \bar{R}}{\Delta h} \approx 2.484$$

Since $\Delta h = 10$, we see that $\Delta \bar{R} \approx 24.84$. That is what we would be willing to pay for the additional hours. ∎

3. Fitting a Model to Data

On p.17 of Chapter 2 we introduced a simple recursion to monitor the value of an investment under a fixed interest rate. Suppose we have an investment which is supposed to be following that model according to a daily interest rate. We'll assume that we don't know the exact interest rate; its existence is a theoretical fact. Suppose we start with \$100, and let M_n be the value of the investment after n days. (Then $M_0 = 100$.) If the daily interest rate is r, then we know that

$$M_1 = (1 + r) \cdot M_0, \quad M_2 = (1 + r) \cdot M_1, \quad M_3 = (1 + r) \cdot M_2, \quad \dots$$

Now suppose we observe the actual value of the investment each day. Here is what we observe over the first few days.

day n :	0	1	2	3	4	5
value M_n :	100	101	100	105	110	111

We want to know the interest rate r. We are supposed to have

$$M_1 = (1 + r) \cdot M_0 \quad \text{which is} \quad 101 = (1 + r) \cdot 100$$

This gives $r = 0.01$. That was easy. But we also want to have

$$M_2 = (1 + r) \cdot M_1 \quad \text{which is} \quad 100 = (1 + r) \cdot 101$$

which solves as $r \approx -0.001$. (Apparently, day 2 was a bad day.) Then again

$$M_3 = (1 + r) \cdot M_2 \quad \text{which is} \quad 105 = (1 + r) \cdot 100$$

and this is $r = 0.05$. There is no one value of r that works in each equation! How are we supposed to find r, given that the observed data don't agree on its value?

Let's back up and realize that we should have expected something like this to happen. Remember that we said that the existence of r is a *theoretical fact* – that probably means that the value of the investment is determined by several things and that the interest rate is an important factor, maybe the most important factor – but perhaps it's not the only factor. In that case, the interest rate wouldn't completely determine the sequence, and so we should not be surprised at the inconsistencies. Here is another possible explanation: suppose that the observed values are only estimates (approximations) of the actual values. There might be errors in those estimates that make the equations inconsistent.

The first sort of error – that theory only approximates fact – suggests that the theoretical r is approximate; the second sort of error – that observations can be rough – suggests that the observations might not reveal the actual r. Either way we are stuck with trying to find a theoretical parameter from inconsistent data.

The situation we have just described is typical of virtually all experiments in both the natural and social sciences. Some kind of underlying theory predicts what should happen, but not exactly; our observations show us what actually happened, but not exactly. Experience shows that even when the theory is rock solid and the observations are as precise as we can make them, some sort of discrepancy between them inevitably occurs. If the theory is important to us, we are still interested in doing the best we can to fit the observed data to the theoretical model. This means getting a kind of *best guess* as to the value of various parameters. We are going to describe how to do this. First, we need to measure a kind of *distance* between two sequences. Second, we describe an optimization problem to find the *best* fit of a theoretical model to observed data. This subject dovetails with statistics, in particular with *regression analysis*, and many of you will see that connection in other courses.

3.1. Square Norm. We need to define a kind of *distance* between two sequences of the same length. Given

$$
\begin{array}{ccccc}
P_0 & P_1 & P_2 & \cdots & P_n \\
\text{and} \quad Q_0 & Q_1 & Q_2 & \cdots & Q_n
\end{array}
$$

we define their *square norm*

$$(P_0 - Q_0)^2 + (P_1 - Q_1)^2 + (P_2 - Q_2)^2 + \cdots + (P_n - Q_n)^2$$

Notice that this distance is a sum of squares – that is to say it is a sum of non-negative numbers. If a sum of non-negative numbers is 0, each of the numbers is 0. In other words, if the square norm is 0, then each term $P_j - Q_j$ is 0, so that $P_j = Q_j$, for each j; the sequences are the same. If the sequences are not the same, then some difference $P_j - Q_j$ is not 0, and the square norm will be positive. The greater the difference between the sequences, the larger will be the square norm.

The word *norm* is a general word used for measures of distance between two objects. The formula for the square norm originates in both geometry and statistics; we will not try to explain how. Here is an example that illustrates the most common use of the square norm.

Example. Going back to our investment example, here is what we observed over the first few days.

$$\begin{array}{ccccccc} \text{day:} & 0 & 1 & 2 & 3 & 4 & 5 \\ \text{value:} & 100 & 101 & 100 & 105 & 110 & 111 \end{array}$$

Suppose we hypothesize a daily interest rate of 1% and we want to measure the discrepancy between that hypothesis and our observation. Here is the 1% sequence; we regard this sequence as *predicted* or *theoretical.*

$$\begin{array}{ccccccc} \text{day:} & 0 & 1 & 2 & 3 & 4 & 5 \\ \text{value:} & 100 & 101 & 102.01 & 103.03 & 104.06 & 105.1 \end{array}$$

Here is the square norm between the predicted sequence and the observed sequence.

$$(100 - 100)^2 + (101 - 101)^2 + (102.01 - 100)^2$$
$$+ (103.03 - 105)^2 + (104.06 - 110)^2 + (105.1 - 111)^2$$
$$\approx 78.01$$

We are not yet in a position to think about the *size* of the norm (78 may seem like a large number). The point is that 78.01 measures the distance between the predicted sequence and the observed sequence. ■

When, as in the previous example, the square norm measures the distance between predicted values and observed values, the norm is often called the *squares error.*

Example. We bring back the Logistic Equation from p.18 of Chapter 2. We had a population sequence P_n governed by the recursive equation

$$P_{n+1} = P_n + k \cdot P_n \cdot \left(1 - \frac{P_n}{E}\right)$$

where the numbers k and E are constants. Suppose we know that $P_0 = 10$, and the claim is made that $k = 0.1$ and $E = 50$. With those values of k, E, here is the predicted sequence.

$$n : \quad 0 \quad 1 \quad 2 \quad 3 \quad 4 \quad 5 \quad 6$$
$$\text{claimed } P_n : \quad 10 \quad 10.8 \quad 11.65 \quad 12.54 \quad 13.48 \quad 14.64 \quad 15.49$$

Now suppose we *observe* the following population sequence:

$$n : \quad 0 \quad 1 \quad 2 \quad 3 \quad 4 \quad 5 \quad 6$$
$$\text{actual } P_n : \quad 10 \quad 11 \quad 12 \quad 13 \quad 13 \quad 15 \quad 16$$

We can compute the squares error (the square norm) between the predicted sequence and the observed sequence to measure the plausibility of the claim about the values of k, E. The squares error is 0.99 in this case. As in the previous example, we are not yet in a position to do anything with that number. ■

Example. We think that the weight w (lbs) of a seven year old boy is predicted by his height h (inches) via the formula $w = 5 \cdot h - 190$. (So, this equation is theoretical.) As an experiment, we record the height and weight of four seven-year-olds in this chart:

$$\text{height}: \quad 47 \quad 49 \quad 45 \quad 46$$
$$\text{observed weight}: \quad 45 \quad 50 \quad 38 \quad 45$$

The equation $w = 5h - 190$ predicts the following weights:

$$\text{height}: \quad 47 \quad 49 \quad 45 \quad 46$$
$$\text{predicted weight}: \quad 45 \quad 55 \quad 35 \quad 40$$

The squares error measures the distance between the observed weights and the theoretical weights: $(0)^2 + (5)^2 + (3)^2 + (5)^2 = 59$. ■

3.2. Least Squares. We have described the situation where we have some sort of observed sequence and a theoretical prediction that it should be explained by particular parameters. In the height/weight example, we had a particular theoretical curve $w = 5 \cdot h - 190$; let's only assume that the curve should be a line $w = m \cdot h + b$ and we'll try to *find* the parameters m, b.

To do this, we imagine picking a particular (but unknown) value of m, b and computing the predicted weights from the given heights:

$$\begin{array}{ccccc} \text{height}: & 47 & 49 & 45 & 46 \\ \text{weight:} & 47m+b & 49m+b & 45m+b & 46m+b \end{array}$$

To test the fitness of a particular m, b, we compute the squares error with the observed data.

$$(47m + b - 45)^2 + (49m + b - 50)^2 + (45m + b - 38)^2 + (46m + b - 45)^2$$

Notice that this is a function of the parameters m, b. Call it $E(m, b)$. To find a best possible m, b, we *minimize* $E(m, b)$. If we are able to do this, the solution m, b is called the *least squares* solution. The least squares solution finds the theoretical model that best fits the observed data.[2]

Minimizing $E(m, b)$ is a non-linear optimization problem with no constraints. One can prove that there is a minimum, and so the First Derivative Test says that $DE(m, b) = \mathbb{O}$ at the solution. The resulting equations (two equations, one for each partial derivative) are called the *normal equations*. We usually allow a numerical solver to find the solution rather than solving the normal equations by hand.

Problem T12.10. Find m, b that minimize the squares error in the height/weight problem.

Solution. We develop numerical formulas to imitate what we just did. Here is what the spreadsheet might look like, showing the row and column labels.

	A	B	C	D	E	F
1	m	b				
2	0	0				
3	h:	47	49	45	46	
4	w(observe):	45	50	38	45	
5	w(predict):	=$A$2*B3+$B$2	⋆	⋆	⋆	
6	delta:	=B4-B5	○	○	○	□

---

[2]See [**4**, Section 8.2].

This is not easy to read; study it carefully! The value of m is stored in A2 and the value of b in B2. The formula in B5 computes $w = m \cdot h + b$ for $h - 47$ in B3. The cells marked $\star$ were obtained by `fill-right` from B5. That's why the references to m, b have dollar signs. Cell B6 holds the difference between the predicted w (in B5, for $h = 47$) and the observed w in B4. The entries in row 6 marked with $\circ$ were obtained by `fill-right` from B6. The objective square norm is in cell F6; the $\square$ symbol stands for this:

$$\texttt{=sumproduct(B6:E6,B6:E6)}$$

We asked the `Solver` to minimize the cell F6, by changing the variables A2:B2 (the parameters m, b), and it came back with

$$m \approx 2.685, \quad b \approx -81.01$$

The minimum squares error was 9.89. (Note: when we invoked the `Solver`, we were careful to *uncheck* the box for non-negative variables, since m, b could be negative.) ■

The line that results from the solution to the previous problem is this:

$$w = 2.685 \cdot h - 81.01$$

This line is called the *line of best fit* or the *least squares line* or the *regression line*. It would be the business of statistics to argue from the minimum squares error back to the plausibility of the theoretical curve. In this course, we will stick with calculating the minimums.

Problem T12.11. An investment of \$100 is supposed to produce a sequence of values M_n with recursion $M_{n+1} = (1+r) \cdot M_n$ where r is a constant parameter. Given the following observed sequence values, find the least squares value of r.

n :	0	1	2	3	4
M_n :	100	130	160	190	230

Solution. We have $M_0 = 100$. Given the number r, we have a *predicted* sequence that satisfies the recursion. We set up such a sequence, using r as a variable, which we initialize to 0. Here is what our spreadsheet formulas look like, showing the row and column labels. Once the formulas are entered, the spreadsheet shows number entries, as usual. This time we lay out our spreadsheet with the data running down a column.

	B	C	D	E	F
1	r	n	M[n]predicted	M[n]observed	del-squared
2	0	0	100	100	=(D2-E2)∧2
3		1	=(1+B$2)*D2	130	=D3-E3
4		2	=(1+B$2)*D3	160	=D4-E4
5		3	=(1+B$2)*D4	190	=D5-E5
6		4	=(1+B$2)*D5	230	=D6-E6
7					⋆

The entry marked ⋆ is `=sumproduct(F2:F6,F2:F6)`. We can use `fill-down` to obtain the computed entries in columns D and F, up to row 6. We have initialized $r = 0$. In the `Solver` we ask to minimize cell F7 (where the squares error is), using the variable B2 (the value of r). The minimum occurs when $r \approx 0.24$. ∎

In the previous problem, suppose that M_0 is regarded as a parameter, as well. In that case, our least squares calculation includes M_0 and r as variables. If you work the problem that way, you obtain a different solution than the one we obtained by working only with r. So, is it *correct* to have M_0 as a parameter? That depends on the purpose of the model within the subject it came from. We are sticking with the mathematics involved.

CHAPTER 13

Spreadsheet Formulas

A *numerical solution* is a solution obtained on a calculator or computer. Numerical solutions always involve approximation, since computed arithmetic is approximate – even if it is pretty accurate. In the Introduction, we mentioned that this course will involve both numerical and analytic calculations. For the numerical calculations, we will often employ Excel.[1] Many campus computers have Excel; of course, you might want your own copy. Among the many online and print resources for learning Excel, we mention the book [7].

The sections of this chapter introduce Excel techniques for performing a wide variety of calculations. We will assume no prior exposure, but even if you are familiar with Excel, you should read this material carefully.

The instructions we give here will assume that Excel is open with the Home tab selected at the very top of the screen. The page before you is called a *spreadsheet*. Its columns are labeled by letters and its rows by numbers. Each rectangle is a *cell*, labeled by the letter and number for its column and row.

1. Function Values

We want to compute the function $y = x^2$ at various values of x.

---

[1]Microsoft Excel. Among the many, many software tools available for numerical work, we have chosen Excel because of its general availability and usage. Apple iWork Numbers and Google Docs Spreadsheet, for instance, have many of the functions of Excel, but they lack the ability to handle the non-linear problems that arise in many applications. For what it's worth, as of 2015 Excel is more flexible and more widely used than its current competitors.

First we explain how to compute a single value of this function. Enter the following symbols and numbers in the indicated cells. You can use the arrow keys or `return` to indicate that you are finished entering something.

	A	B	C	...
1	x	y		
2	4	=A2*A2		
⋮				

The expression =A2*A2 is a *formula*. The equals sign at the beginning tells Excel to compute the value of this expression rather than to display the formula itself. The A2 in the formula refers to the contents of that cell. Since 4 is in cell A2, cell B2 shows 16 when the formula is entered. If you select B2 and look at the formula strip just above the spreadsheet grid, you will see the formula.

Change cell A2 to various other values, and notice that B2 shows the square of A2 each time.

When you are typing in the formula, you can click on A2 instead of typing it. Try it by re-entering =A2*A2.

The expression A2*A2 could also be written A2∧2 The symbol ∧ is exponentiation. Try changing B2 to =A2∧2 and make sure you are still computing the square after the formula is entered.

Let's compute a list of function values, using a more complicated function. We'll use $y = x^3 \cdot \exp(-\sqrt{x})$. Change B2 to

$$=(A2\wedge 3)*\exp(-\text{sqrt}(A2))$$

Notice how x^3 is denoted; `exp` is the exponential function and `sqrt` is the square root function. For the square root, we could also have written

$$A2\wedge 0.5 \quad \text{or} \quad A2\wedge(1/2)$$

Put the following values in cells A2,A3,A4

	A	B	C	...
1	x	y		
2	4	=(A2∧3)*exp(-(sqrt(A2))		
3	3			
4	1			

Cell B2 should have 8.66.... We want to compute $x^3 \exp(-\sqrt{x})$ for the values 3 and 1, as well. We can do this *without* rewriting the formula over and over. Select B2 and place the cursor over the small square at the lower right corner of that cell. The cursor should turn into a solid cross. (Not a double-lined cross and not the hand symbol.) Once you have the solid cross, click and hold down the cursor. Drag the cursor down to the cells B3,B4 below B2 and let go! Here is what you should see.

	A	B	C	...
1	x	y		
2	4	8.661		
3	3	4.776		
4	1	0.367		

(You may have more decimal places.) Click on cell B3, and look at the formula strip above the spreadsheet: you will see that the formula that was in B2 has been copied down, but the reference to A2 has been changed to A3. The operation of pulling down with the solid cross is called a `fill-down`.

We will understand `fill-down` by using it, but here is an abstract description: when you `fill-down`, the cell row references (these references are numbers) are increased by one for each row drag down the column.

Let's get a list of function values where the function has a parameter. Let $y = \ln(m \cdot x - 2)$, where m is a parameter (constant). We want to compute y when $x = 1, 2, 3, 4$. We'll start with $m = 10$; later we will change it. Set up the following cell contents. (Notice that we have started in cell E4; we're just making the point that the fun can start anywhere!)

	E	F	G	H
4	m	10		
5	x	y		
6	1	=ln(F$4*E6-2)		
7	2			
8	3			
9	4			

In the entry F$4, ignore the dollar sign temporarily; that refers to cell F4, which holds the value of m. When you enter the formula in F6, the number 2.079 is displayed; that's the y-value when $x = 1$ (and $m = 10$).

Use fill-down from cell F6 to F9 to get the other y-values for $m = 10$. Then look at the formulas in those cells. The E-column changes, row by row – running over the four different x-values. But notice that the reference F$4 does not change. That's the purpose of the dollar sign: to keep row 4 constant as the fill-down is done.

Now we want to change m from 10 to 11, and then to 12. One way to do this is simply to change the entry in F4 to 11. Try that, and notice that all the y-values change as well! The y-values update to the new value of m.

We want to show how to change m while keeping the previous values of m around – say to study what happens as m changes. Change F4 back to 10 and enter 11 in G4 and 12 in H4.

Next we need to change the formula in F6; change it to this:

$$=\ln(\text{F\$4*\$E6-2})$$

We will explain the dollar sign on E momentarily. Then, fill-down from F6 to F9.

Now click on F6 and shift-click on F9; the cells F6,F7,F8,F9 should all be highlighted. Place the cursor over the small square at the lower right corner of the box of cells, hold down, and drag to the right, so that the box extends to the right. End up at H9. Now release the cursor, and all the cells between

G6 and H9 should fill in with numbers. This is called a `fill-right`.

Click on G7 and look at the formula strip. You should see `=ln(G$4*$E6-2)`. Notice that the m-value referred to is G$4. As you `fill-right`, the column references change but the row references stay the same. The F in F$4 did not have a dollar sign, and so it changed to G in the fill right. If the F had had a dollar sign $F$4, it would not have changed. On the other hand, the x-value $E6 does have a dollar sign, and so that column stayed the same in the fill right. Each column refers to the same list of x-values in column E, but they refer to the m-value at the head of their particular column.

To summarize: we have introduced formulas to calculate function values, `fill-down` and `fill-right` as ways of replicating formulas, and the dollar sign to prevent a fill operation from changing a particular cell reference. We will get lots of practice with these operations as we use Excel throughout the course.

2. Recursion Calculations

Let's see how to compute a recursive sequence. We'll start with a very simple example:

$$Q_{n+1} = 1.5 \cdot Q_n \quad \text{for} \quad n = 0, 1, 2, \ldots$$

We are going to compute the values of Q in column B, starting with Q_0 in B1, and then Q_1 in B2, and then Q_2 in B3, and so on. Let's have $Q_0 = 2$; enter 2 in B1. In B2 we want Q_1, and the recursive equation is $Q_1 = 1.5 \cdot Q_0$. Enforce this via the following formula in B2:

$$=1.5*B1$$

Now B2 shows $3 = 1.5 \cdot 2$, which is Q_1. For the rest of the sequence, we simply `fill-down`: select B2, get the solid cross in the lower right corner, and drag down to B10. Cell B10 should show 76.88, or so. (That's Q_9.)

Suppose we wish to add up the Q_n, for $n = 0, 1, \ldots, 9$. These values are in cells B1 through B10; we refer to this block of cells as B1:B10. In cell B12, write the formula =sum(B1:B10); the sum of the terms should appear: 226.66, or so.

Let's do the same formula a different way. Select B13, and type

$$=\text{sum}($$

Now click on cell B1, and then shift-click on cell B10. Notice that B1:B10 appears in the formula strip. Type a right parenthesis and press return, and you should have the same formula as in B12.

Here is a more complicated set-up that will be discussed in class. The recursion involves parameters x, y:

$$F_{n+2} = x \cdot F_{n+1} + y \cdot F_n \quad \text{for} \quad n = 0, 1, 2, \ldots$$

We let $x = -1$ and $y = 2$ and $F_0 = 3$ and $F_1 = 7$, but we want to be able to change these values.

	A	B	C
1	x	-1	
2	y	2	
3	n	F[n]	
4	0	3	
5	1	7	
6	=A5+1	=B$1*B5+B$2*B4	
7			

Select A6 and shift click on B6, so that A6:B6 is selected. Now fill-down to get more values of n and more terms of the sequence. The references B1 and B2 to x, y, respectively, will not change. You should notice how the recursion updates as you go down.

To change x, y, we can simply change them in cells B1,B2. If we want to keep the previous values of x, y, we can put new values in column C, and use fill right from row 4 down to update the sequence.

3. Matrix Calculations

In Excel, the rectangle of cells with upper left corner at A1 and lower right corner at C4 is referred to as A1:C4. This block of cells will hold a 4×3 matrix, as you can see by selecting that range on a spreadsheet.

Excel matrix addition is straightforward. Suppose we want to compute the sum of two 2×3 matrices, one in block A2:C3 and one in block D4:F5 We select a block of 2×3 cells in which to store the answer: say B7:D8. In the upper left cell B7 we type the formula

$$=A2:C3+B7:D8$$

What you do next depends on the type of your computer. On a PC running Windows[2], press CONTROL+SHIFT+ENTER; on a Mac[3], press APPLE+RETURN. We call this kind of return a *matrix return*. The matrix return will fill the block with the product sum. Typing an ordinary RETURN will not give you the entire matrix sum.

Scalar multiplication is similarly easy. Say we want to compute $A - 3 \cdot B$ where A, B are 3×4. Assume that A is in cells A1:D3 and B is in F1:I3. If we want to put the answer in B5:E7, then we select those cells, and type the following formula in B5

$$=A1:D3-3*F1:I3$$

Do a *matrix return* to fill B5:E7 with the answer.

The formula mmult multiplies matrices. Here is an example: we want to multiply the 3×4 matrix in A1:D3 by the 4×3 matrix in F1:H4.

---

[2]Microsoft Windows
[3]Apple Macintosh, iMac, MacAir, etc.

	A	B	C	D	E	F	G	H
1	1	0	-1	2		1	3	-1
2	1	1	4	6		0	1	5
3	0	2	-3	4		-2	3	-2
4						1	0	9

Now we compute the product. It doesn't matter where it goes, but we need room for the product matrix. Our product is 3×3, so maybe we move down a ways, starting in row 11. We select a 3×3 block of cells, starting with the upper left cell in the block. Say we select A11:C13. Here is formula typed.

	A	B	C
11	=mmult(A1:D3,F1:H4)		
12			
13			

Don't forget to do a matrix return after typing the mmult formula. The matrix product should appear.

	A	B	C
11	5	0	9
12	-1	16	50
13	10	-7	52

To compute the inverse of a matrix, you use the minverse formula. Here is a 2×2 example.

	A	B	C	D	E
1	2	-4		=minverse(A1:B2)	
2	7	6			

The minverse formula is typed while D1:E2 is selected, using a matrix return, as you would expect. If the matrix that you try to invert does not have an inverse, each cell in the selected area will display #NUM!, indicating an error.

The mdeterm formula determines whether a square matrix has an inverse. (Recall that a non-square matrix cannot have an inverse.) This formula gives a single number, the *determinant* of the matrix. A square matrix has an inverse if and only if its determinant is not 0. In the previous example, we could compute `=mdeterm(A1:B2)`. This formula computes a number, so it is entered with an ordinary return. This matrix has determinant 40.

An important note: You cannot edit any entry in a matrix computed from a matrix return. If you do that by mistake, you will get an error message; you will have to cancel the formula by clicking the delete-x on the formula text field.

4. The Solver

The `Solver` is an *optional add-in* in Excel. If it does not appear on the `Tools` menu, it should be available to be downloaded through the `Tools` options. The `Solver` is set up to solve optimization problems, but it can be used to solve linear equations, as well. We will demonstrate its use in class. There are many online examples and tutorials.[4]

---

[4]For instance, see http://www.solver.com/excel-solver-help

Bibliography

Textbooks for applied calculus

[1] Raymond A. Barnett, Michael R. Ziegler, Karl E. Byleen, *College Mathematics for Business, Economics, Life Sciences, and Social Sciences*, 12th edition, Prentice Hall, 2011.

[2] Geoffrey Berresford, Andrew Rockett, *Applied Calculus*, 6th edition, Brooks/Cole, 2013.

[3] Larry J. Goldstein, David C. Lay, David I. Schneider, Nakhle Asmar, *Calculus and Its Applications*, 11th ed., Prentice-Hall, 2007.

Applications of this course to various disciplines

[4] Elizabeth S. Allman and John A. Rhodes, *Mathematical Models in Biology*, Cambridge University Press, 2004.

[5] Robert Dorfman, Paul A. Samuelson, and Robert M. Solow, *Linear Programming and Economic Analysis*, New York: McGraw-Hill, 1958, p.283ff.

[6] Frank C. Hoppensteadt and Charles Peskin, *Mathematics in Medicine and the Life Sciences*, New York: Springer-Verlag, 1992.

[7] Matthew MacDonald, *Excel 2013: The Missing Manual*, O'Reilly Media, 2013.

[8] J.D. Murray, *Mathematical Biology*, New York: Springer, 1989.

[9] Raymond A. Serway and John W. Jewett, Jr. *Physics for Scientists and Engineers with Modern Physics*, 7th edition, Thomson Brooks/Cole, 2008.

[10] Knut Sydsaeter and Peter J. Hammond, *Essential Mathematics for Economic Analysis*, London: Pearson Education Limited, 2008.

Technical mathematics related to applied calculus

[11] Carl B. Boyer, *The History of the Calculus and its Conceptual Development*, New York: Dover Publications, 1959.

[12] George B. Dantzig, *Linear Programming and Extensions*, Princeton: Princeton University Press, 1963.

[13] C. Henry Edwards and David E. Penney, *Calculus*, 6th edition, Pearson, 2002.

[14] Sir Thomas L. Heath, *A Manual of Greek Mathematics*, Mineola, New York: Dover Publications, 2009.

[15] D.E. Smith, *History of Mathematics*, Volume 1, New York: Dover Publications, 1923.

[16] Alan Tucker, *Applied Combinatorics*, 6th edition, John Wiley and Sons, 2012.

Index

$(a + b)^2$ formula, 31
$A[i, j]$, the ij entry of matrix A, 145
A^{-1}, inverse of matrix A, 175
$F\big|_a^b$, difference notation, 107
I_n, the $n \times n$ identity matrix, 151
$[a, b]$, closed interval, 107
Δ, change, 27
$\approx$, used for all numerical answers, 5
exp, the exponential function (base e), 2
$\frac{\partial}{\partial x}$, partial derivative with respect to x, 189
$\int f(x) \cdot dx$, the indefinite integral, an antiderivative, 100
$\int$, the integral, 100
$\int_a^b f(x) \cdot dx$, the definite integral, 107
$\lim_{x \to a}$, the limit as x goes to a, considered informally, 31
$\ln(x)$, the natural logarithm (base e), 3
$\mathbb{O}$, the zero matrix of various sizes, 146
$\mathcal{R}$, Riemann sum, 113
∂, del, 189
e, base of the exponential function, 1
$f''(x)$, the second derivative of f, 69
$f'(x)$, the derivative of $f(x)$, 38
$n \to \infty$, n gets large without bound, 113

acceleration, 36, 42
Addition Rule for derivatives, 46
amp, abbrev. ampere, 9
ampere, 9, 182

analysis, mathematical, v
antiderivative, 99
area, as an application the integral, 120
augmented matrix, 165
average density, 34
average velocity, 30

Babylonian method for square roots, 23
basic variables, 171
Binomial Theorem, 45
block of cells in Excel, 233
bounded constraints, 208

calculus, v
capacitor, 9
cell, in an Excel spreadsheet, 228
cell, in Excel, 227
Chain Rule, 56
Chain Rule, differential version, 75
Chain Rule, for functions of several variables, 197
closed constraint, 205
closed interval, 107
Cobb-Douglas model, 10
coefficient matrix, 164
composite function, 56
compounded, 17
concave down, 70
concave up, 69
concavity, of a curve, 69
consistent system of equations, 171

239